Enamel-working in Iron Age, Roman and Sub-Roman Britain

The Products and Techniques

J. D. Bateson

BAR British Series 93
1981

B.A.R.

B.A.R., 122 Banbury Road, Oxford OX2 7BP, England

B.A.R. 93, 1981: "Enamel-working in Iron Age, Roman and Sub-Roman Britain"

ISBN 9780860541448 paperback

ISBN 9781407323695 e-book

DOI https://doi.org/10.30861/9780860541448

A catalogue record for this book is available from the British Library

This book is available at www.barpublishing.com

CONTENTS

LIST OF FIGURES

PREFACE

Enamelled objects are relatively common both as site-finds and display objects in many museums. Enamelwork is an ancient technique well known to archaeologists. The Battersea shield, dragonesque brooches, enamelled cups or hanging bowl escutcheons at once spring to mind. However when it comes to the question of what exactly the enamel is, how it was made and applied, and to which objects, when and where, answers tend to be vague and uninformative. The Belgae, champleve and the Villa D'Anthee tend to figure largely in replies yet they mean little or have now been discredited. The only major work on early British enamels is now over forty years old (Henry, 1933, pp. 66-146). Beyond this only a few brief paragraphs, mainly on parts rather than the whole, and more recently two general articles (Butcher, 1976, pp. 42-51 and 1977, pp. 41-69) can be found on the subject.

Enamelwork would therefore seem to be a topic worth consideration in as many aspects as possible. It is however a wide field. This study concentrates on the enamelwork of Roman Britain but in fact covers the period from 200 B.C. to A.D. 700. It would be difficult to pick up the story in A.D. 43 and to drop it in A.D. 410; it is necessary to look at what went on before and after the Roman occupation. Over this rather long time-span the enamelled objects made in Britain are comprehensively surveyed. A visual examination is complemented by a survey of the available scientific analyses and the results of some experiments. The technical aspects of the craft are dealt with under three headings: the vitreous substance, the metal base and the methods of joining the two together; millefiori enamels receive particular attention. A new scheme for describing Romano-British enamels is suggested and possible manufacturing sites in Britain are listed. The study concludes with a picture of the repertoire and techniques of the Romano-British enamel-working industry, its periods and areas of manufacture, the effects of imports, its markets and set-up.

In the end one should be able to say more about an enamelled object other than colour and a bland assertion that it is a champleve enamel of the second century. One may say more precisely just what the 'Barbarians of the Ocean' were up to.

This report is based on a thesis presented at The Queen's University of Belfast in 1977. I am grateful to my supervisor Professor Martyn Jope for suggesting the subject and for his guidance and practical assistance. More recently I have had much help from Dr. Grace Simpson of Oxford who is currently editing the late M.R. Hull's corpus of Roman brooches for a subsequent report in this series; our conversations on the brooches have been most useful and the manuscript has generously been made available. I would also like to express my sincere gratitude to the staff of the museums listed in note 4. Without exception I received the fullest co-operation and help and only with this access to the basic material could the research have been carried out. Finally mention must be made of friends and colleagues, too numerous to list, who kindly brought objects to my attention or discussed various points with me.

The scale on the figures is 1:1 except where otherwise stated or where an object as such is not being illustrated e.g. the millefiori patterns of figures 12 and 13. Figure 1 is a colour code used throughout the rest of the figures. The illustrations are intended to show types or techniques, often with the enamel either omitted or shown complete, and should not be taken as representing specific objects.

INTRODUCTION

Enamel is generally described as a vitreous substance fused on to metal. It can also be applied to glass and though true enamelled glass does not occur in Roman contexts, some of the glass beads and bangles with inlaid spirals or bands such as were found at Traprain Law are sometimes described as being enamelled.[1] The present work however adheres to the strict definition which deals only with a metal base. It may also be as well to mention that niello is not included. Surprisingly either through ignorance or because it resembles enamel, niello is not always distinguished from enamel. However the basis of the former is metallic[2] and could not be included with a substance, the basis of which is vitreous.

The technique is looked at only in Britain and not in Ireland. The huge number of enamelled objects of the Roman period found in Britain are absent from the corresponding period in Ireland while the large number of enamelled objects belonging to the Early Christian Period in Ireland are not matched in post-Roman Britain. Ireland will not however be entirely ignored since the two major workshops at Garranes and Lagore provide useful comparative material.

All enamelled objects found in Britain are dealt with. This means that those of Continental origin are included to some extent. This has made it necessary to look briefly at the Continental background so as to be able to distinguish between those objects which are of British manufacture and those which are imports from the workshops of Gaul or the Rhineland. One may then see more clearly just which are British products and how they were influenced by the imported material. Equally one may grasp some idea of exports to the Continent.

The primary purpose of the project is to examine the enamels of the Romano-British period. However the enamelwork of the Iron Age and early post-Roman periods is also covered, though not so fully. The time span is therefore approximately 200 B.C. to A.D. 700 with greatest emphasis placed on the first four centuries of the Christian era. The Roman enamels cannot be looked at in isolation though the great majority of objects examined do belong to the Roman period. However the Romans did not bring enamel-working to Britain. It was here when they arrived, it continued after they had departed. When talking of Romano-British enamels one means a British atelier in a Roman mileau for enamelwork was a native contribution to the Roman world. There appears to be no break between the pre-Roman and Roman enamellers in Britain: products may change but seemingly not the makers. It has been suggested (Henry, 1933, p.146) that there was a break between the Romano-British industry finishing in the mid-third century and the makers of the enamelled escutcheons of the hanging bowls of the early Anglo-Saxon period. It is argued later that no such break took place and that the enamels of the post-Roman epoch are in fact the tail end of an industry giving a last kick in the revived native atmosphere. This is not to deny that the flowering of the Romano-British enamel-working industry took place in the second century.

It is proposed to look closely at the manufacturing process from start to finish that is from the raw materials used for the enamel itself through to the final cleaning of the end product. Emphasis is placed on the enamel, the metal backing and the various ways of making the former adhere to the latter. This is done by means of a visual examination of the enamelled objects, by comparison with modern enamelling techniques, by experiment and by examining the available analyses of enamels and metals. It should be pointed out perhaps that the study is of techniques rather than a scientific explanation of the actual chemical processes.

The basis of this work is therefore an examination of the enamelled objects themselves. The majority of those found in Britain are looked at whether or not they still retain their enamel. There is such a wealth of variety in the types of objects enamelled that it would be attempting too much to carry out a detailed study of each type or to produce a complete corpus of all enamelwork ever found.[3] The enamelled brooches or even a portion of the brooches would probably suffice as a study but to do that would be to look at only part of the enamel-working story. This however is something which should be dealt with *in toto*. Thus while dealing with a type of enamelled object no claim is made that the type is completely looked at. Nevertheless since many of the types are lacking such a study, that included here may be of some value beyond the enamel aspect alone. On the other hand it should be pointed out that in the study of other groups the unenamelled specimens are not included e.g. only the enamelled head stud brooches or trumpet brooches are being discussed and the implications may not apply to the non-enamelled specimens. In all about two thousand objects are covered. Those of the Roman period are divided into three main groups: bow brooches, plate brooches and non-brooch material. The objects of the Iron Age mainly consist of pieces of horse harness while in the early Saxon period one is mainly concerned with the escutcheons of hanging bowls though in both cases this is an over-simplification.

The search for material has been fairly comprehensive. No claim is made however that every enamelled object found in Britain has been examined, nor is this necessary since overall patterns are being dealt with and not one closely defined group of objects. Nevertheless the majority of the material available has been looked at. In those few instances where a list of specific objects has been published it has been possible to check how complete the search has been. Thus of the group S iii brooches (Richardson, 1960, pp. 200-213) or the dragonesque brooches (Feachem, 1951, pp. 32-44 and 1968, pp. 100-102) eighty per cent of the available material has been examined. Equally in 1954 (Hildyard, 1954, p. 73) fifty-four group Q or head-stud brooches were cited including forty-one with enamel; for the purpose of this study over one hundred and fifty were noted. Again ninety enamelled specimens of the trumpet brooch group are now dealt with as against twenty-one included by Hildyard (*op. cit.*). The major national museums in London, Edinburgh and Cardiff along with the more important collections such as those at Oxford, Bristol and York have all been checked for relevant material which was then carefully scrutinised. Likewise the more important site collections such as that from Silchester, housed at Reading, and that from Caerwent, at Newport and those still on the actual sites as Caerleon, Chesters and Corbridge were examined. A full list of all museums visited is given in note 4; no major collection has been omitted. As regards distribution of provenances, types of sites and variety of object there appears to be adequate coverage.

General collections are housed at the British Museum and the Ashmolean Museum; the latter also has a useful number of objects from the Continent thus providing some non-British material which was examined closely. The National Museum of Antiquities of Scotland provides an excellent range of material from Scotland whether it be a native site such as Traprain Law or a Roman fort such as Newstead. The National Museum of Wales in Cardiff similarly deals with Wales. The Wall area has yielded many enamelled objects to be seen from Carlisle to Newcastle-upon-Tyne and the site museums between the two. Civic centres are provided for by the Museum of London and others such as Colchester, Saint Albans and Cirencester. Overall there is a fairly representative selection of objects available for study. A fair proportion of these remain unpublished.

On the other hand much has been gleaned from publications. Those checked are listed in note 5. The complete runs of the major national journals have been fully perused with good results. Likewise the reports of the Research Committee of the Society of Antiquaries of London have proved to be of great value. Those books which have produced material quoted will be found in the Bibliography. Many were also checked which yielded no enamelwork. As regards the Continental enamelwork greater reliance has had to be placed on the written work but fortunately a large number of objects are listed in such works as Henry (1933, pp. 66-146), Exner (1939, pp. 33-121) and Sellye (1939, pp. 1-91).

It may be as well to mention some caveats at this stage. If an enamel has retained its colour there is no problem but once decay has set in difficulties arise as to what the original colour was and here subjectivity can play a great part. When an enamel decays it generally turns a whitish or greenish colour and frequently an enamel which is now in such a state is described as being 'red' meaning a decayed or weathered red though writers frequently omit the word 'decayed'. In many cases this may be a correct description but is not always so. Green and blue can go the same way; several semi-decayed examples exist. Often too what is described as 'green' is in fact another colour in a state of decay.

Apart from decay it is not always possible to give an exact colour to an enamel. Yellow merges into orange which sometimes merges into red. Greens merge with blue and light or watery blues merge into white. Dark blues merge into black. It is surprising how often the description of a colour given in a museum register differs from that given in the published account. This report tries to give only <u>certain</u> colours in its descriptions; uncertain ones are preceded by by a question mark and less certain ones are described as decayed.

Objects from which enamel has apparently disappeared are included. In some instances, particularly with the more unusual type of object, it could be argued that the piece never held enamel at all. Equally the theory that the setting is intended for a stone could be advanced. On the whole though with objects which no longer have enamel it is usual that they belong to types which do have it. In these cases it is also possible to say with some certainty what the actual colours would have been. However it could also be that the empty cells never contained enamel because the object was never finished for some reason. One cannot ever be really sure in such a case. There are some clear examples of unfinished products which are dealt with in the appropriate place. Objects which have lost their enamels are useful as a pointer to the rate of decay, poor workmanship (through the inability of the enamel to adhere) and the method of pre-treating the cells. This makes up for the many examples where the enamel is complete and one cannot be sure of what the position is underneath.

Finally there are two further drawbacks in relation to museum material. The lesser is that on some objects where the enamel has disappeared it has been 'restored' usually in red wax (on the whole the colour appears to be correct). The other great problem with the museum objects is a lack of a stratified provenance such as with the old finds from York or many of the stray objects from Leicester. This is even the case with some finds from excavations, some of them carried out not all that long ago. While the actual object is of primary concern nevertheless more stratified finds would be useful. However the position appears to be better than in Pannonia where there appear to be only two well-dated pieces of enamelwork (Sellye, 1939, p. 34).

Something should now be said about terminology. Such words as champleve and cloisonne are used far too often without a proper understanding of their meaning. They are words properly applicable to the enamelwork of the last few centuries. There are other lesser known techniques used by modern enamellers, the words for which could be applied to some Romano-British enamels and techniques used only by earlier enamellers for which there is no modern equivalent. When later it emerges that all Romano-British enamels are champleve enamels it becomes obvious that the word has little meaning in this period. However it is used where relevant since it is too well-known and useful to be discarded entirely. 'Millefiori' and 'mosaic' are both used indiscriminately in many publications to denote millefiori enamel - a term which needs (and receives) proper definition. The word mosaic is also used at times to denote just different coloured enamels on the one object. Frequently the word cloisson is used for the metal walls of cells even where the technique used is champleve and not cloisonne. Here it is hoped to amend such abusage and to be explicit in the use of terms in the text. A short glossary is added to help make the matter clearer.

Mention should also be made of references to enamelwork in ancient writings. However in comparison to glass for instance this line of enquiry is very unrewarding. The only helpful reference is the oft quoted statement of Philostratus concerning the enamelled horse trappings of the barbarians living by the Ocean.[6] Other than this there are only a few and vague references noted elsewhere (e.g. Thurnan, 1871, p. 482).

Before turning to a detailed study of the objects it may be useful to see what modern writers have to say on the subject.

No previous study of enamelwork is comprehensive in either covering the whole period defined above or even one of the parts, the Iron Age or post-Roman or more particularily the Romano-British period. The best pieces on Iron Age enamelwork are the few pages by Leeds (1933, pp. 39-48 and passim) and Smith (1928, pp. 101-106). These are no more than exceedingly general views of the subject. Fox (1958, passim) lists many of the enamelled objects in his magisterial work on Iron Age art but has little to say on enamel as such. The major Iron Age hoards - a valuable source of information - have only recently begun to receive proper attention though the enamel decoration remains poorly described. The recent work on the actual red glass enamel (Hughes, 1972, pp. 98-107) however sheds some light on the enamel itself although the paper raises as many problems as it solves. Part of the Iron Age in Britain and Europe is covered by Francois Henry (1933) in her article which deals mostly with Roman Britain: however it is not comprehensive; the objects are not well described; it ends rather suddenly in the mid-third century and argues that no millefiori enamel was made in Britain - believing most to have been manufactured at the Villa D'Anthee; there is very little on the technical aspects of the subject; however a fair amount of European enamelwork is included. The latter overall has been somewhat better examined than the British with Exner (1939) looking at the Rhineland brooches, Sellye (1939) dealing with the Pannonian enamelled finds and more recently Dr. Spitaels examining the Belgian enamelled brooches.[7] No such work exists for Britain though individual types are to some extent looked at but without specific reference to their enamel decoration. All the brooches have been arranged by Collingwood (Collingwood and Richmond, 1969, pp. 286-303) though in a simple fashion and without regard to any enamel which may be on them. The arrangement has proved its use over the years and is followed here. However it is very simple in places e.g. the plate brooches have had to be expanded. Hull (forthcoming) provides a full corpus but concentrates on typology and has little to say on the enamel as such.[8] Drogonesque brooches (Feachem, 1951 and 1968) and Collingwood's sub-group Siii (Richardson, 1960) have been examined as groups while among non-brooch enamelwork only the massive armlets (Simpson, 1968, pp. 233-254) and the button-and-loop fasteners (Wild, 1970, pp. 137-155) can be said to have been examined in depth. More recently Miss Butcher has contributed a valuable article on Romano-British enamels (Butcher, 1976, pp. 42-51) and followed this up with the detailed publication of a group of enamelled objects recently recovered from various sites in Britain (Butcher, 1977, pp. 41-69) while the miniature stools (Green, 1975, pp. 54-70) and enamelled cups/paterae (Moore, 1978, pp. 319-326) have received more cursory attention. From the post-Roman period hanging bowls have of course had more written about them than enough and yet an elementary corpus is still lacking. One of the few angles they have not been examined from is that of the enamelwork. Since enamelled hanging bowl fittings constitute a large part of the post-Roman body of enamelwork an examination of them seems desirable. Overall then a comprehensive study of enamelled objects and their techniques is sadly lacking.

An attempt is made here to remedy this situation. The work is divided into two major parts: the first dealing with the enamelled objects themselves and the second with the process of enamel-working. The former tries to answer the question, what objects were being enamelled in Britain during the period previously defined. The major part of this deals with the enamelled objects of Roman Britain, that is to say the brooches first and then what may be termed non-brooch material. Both bow brooches and plate brooches are examined and an attempt is made to define and list these somewhat better than they have been previously. Non-brooch material covers seal-box lids, miniature stools, mounts, belt-plates, cups and other objects many of which require investigation in their own right. The chapters on the Romano-British objects are preceded by one on the Iron Age enamels since to a large extent 'this is where it all began'. Likewise after the Romano-British enamelwork there is a chapter on the post-Roman enamelwork.

Once the 'evidence' has been dealt with we move on to the techniques, which constitute the second part. Using the definition of an enamel as a vitreous substance applied to a metal base and fused to it by the application of heat, the three major chapters cover the actual enamel (the vitreous substance), the metal base, and finally the methods of applying the former to the latter. Since we are looking at a subject on which so little has been done there are not many specific questions which may be asked. For the most part we must stick to the general question of what is it all about? How did the Romano-British craftsmen make their enamels? What exactly is the enamel? What bases are used? How is the enamel applied? The examination of the objects shows that in the application of the enamel to its base three techniques are used which may be called simple, complex and millefiori enamels. With the latter we can be more specific in our questions. Is there such a thing as a millefiori enamel? What exactly is it? Was it ever made in Britain? The remaining chapter, 9, looks critically at certain sites where enamel-working is alleged to have been carried on and then tries to draw up a list of sites in Britain where the process actually appears to have occurred.

Finally we look at the overall picture to see what conclusions can be drawn from the previous discussions. Then it is hoped that a clearer picture of the Romano-British enamel-working industry will have emerged.

PART I THE PRODUCTS

CHAPTER 1

THE ENAMELWORK OF THE IRON AGE

Iron Age enamels are examined first since Romano-British enamelwork cannot be studied in isolation. The art of enamelling was not carried to Britain by the invaders in A.D. 43. When the Romans arrived they found an already flourishing enamel-working industry which continued to function not only during their occupation but long after they had left. The Romano-British industry cannot therefore be easily compartmentalised. To examine it properly and to understand it the preceding Iron Age enamels must be examined too.

It is exceedingly difficult to draw the line between the two industries. As far as one can tell the same native enamellers made terrets or the like for their chieftains and then produced quite different objects for their new patrons. Again the transition from an Iron Age cultural background to a Roman one did not occur all at once throughout Britain nor indeed did it happen particularly rapidly. In the frontier or military zones this was more the case. Hence a basically Iron Age terret could have been made well into the Roman period, as late as the second century A.D. Thus as regards the Iron Age enamels there are included objects basically Iron Age in style and type but manufactured in the post-Iron Age period. Most fit into their category quite well but some, such as button-and-loop fasteners, pose problems. These are not what one would call a Roman type of object though most were probably made well into the Roman period. Here the more obvious Iron Age examples are included in this chapter while leaving the bulk to Chapter Four. Roman is perhaps a bad term - Romano-British is better and perhaps for this chapter British enamels.

One problem should be mentioned now. Many of these objects are well known through their inclusion in books and articles on Iron Age art. Yet good descriptions of the articles and more especially of their techniques are relatively few. This applies as much to the hoards as to many of the individual objects. However this study concentrates on the enamel as well as the backing material and the method of attaching the one to the other.

There are approximately three hundred objects which are products of the British Iron Age and which bear enamel decoration. These consist of just under two hundred individual finds - many of them the finest pieces of Celtic art from this country - and around one hundred pieces which constitute parts or all of seven of the major Iron Age hoards viz. those from Middlebie, Polden Hill, Saham Toney, Santon, Seven Sisters, Stanwick and Westhall. The objects (with reference to the hoard specimens where appropriate) are dealt with by types and then the hoards individually. There are three basic groups: harness/chariot equipment, military equipment, and miscellaneous types ranging from fibulae to mirrors.

Harness and chariot equipment

This is by far the largest group and within it terrets are the most numerous. There are over eighty examples with enamel decoration and of these about half were found in the hoards mentioned above. The only typology of terrets available is over thirty years old (Leeds, 1933, pp. 118-126) but is still useful, although some of the more recent finds do not easily fit into it.[9] Leeds divided the terrets into eight groups of which numbers 1, 7 and 8 bear no enamel while the remaining five groups include enamelled examples. Those of group 2 are invariably enamelled on their broad faces with either large curving cells containing red enamel only such as that from Runnymede[10] or red enamel with the addition of separate spots of blue or yellow enamel as found on the 'Suffolk'[11] and Auchendolly[12] specimens respectively. Examples with the enamel in smaller, more simple cells also occur.[13] Only some of groups 3 and 4 - both lipped terrets with the lips at right angles to the ring in group 3 and parallel to it in group 4 - are enamelled. On these the enamel - always red - is found on the lips[14] and on the sides of the ring. Several examples of each group were found in the hoards from Stanwick and Polden Hill respectively (see under Hoards below). Groups 5 and 6 are again basically similar with three knobs set on the edge of the ring. Those on group 5 are round or flat-topped with a quatrefoil design in enamel while those of group 6 have rectangular or oval knobs. Red, blue and yellow are the recurring colours of the enamel usually one of the latter two combined with red as on the group 5 specimen from Brampton (Green, 1972, p. 346, fig. 2) with red and blue enamel in small geometric cells or the group 6 terret from Traprain Law[15] with small square cells of red and yellow. The example from the Cambridgeshire Fens has blue, yellow and another colour - now decayed but probably red - on its knobs, as well as the sides and faces of the ring.[16] On the whole most of the enamelled terrets can be fitted into these five groups.

There are however three interesting and important terrets which do not quite fit Leeds' system. They are worth mentioning in some detail. All three are similar with broad faces and an open square centre to the bar (with an iron bar in the centre). Enamel decoration is found on the faces of the ring, where there is a design of hooked scrolls, as well as on the faces of the bar centre. The first comes from London (Borrajo, 1908, pl.V, no. 2) and still bears slight traces of red enamel on one of the faces; the bar faces here are divided into crude triangles, now empty. The other two come from abroad. One from Eauze, in the south of France had red enamel (Michon, 1925, pp. 145-7), while the other from the Fayum in Egypt has both red and blue enamel on the faces of the ring (Smith 1925, p. 87, fig. 81). Although these broad enamelled faces are a feature of group 2 terrets the treatment of the bar is akin to that found in group 8. However it has been suggested that these three objects might have been manufactured on the Continent rather than in Britain (Spratling, 1980, p. 116).

Meanwhile there is not much that can be said about the next item of harness furniture - the cheek-piece - of which fourteen enamelled specimens are known. As a whole bronze cheek-pieces are not common in the Iron Age (Piggott, 1952, p. 21). Eleven of the specimens are of normal type and six of these come from the Polden Hill hoard.

Some have only red enamel as that from Eckford[17] while others have two colours, the second usually decayed but on the Bower Chalke specimen blue (Leeds, 1933, p. 103). The three odd pieces are S-shaped and are indeed probably not cheek-pieces at all. Fox (1958, p. 130) calls them dolphin-shaped as the two specimens from Polden Hill where missing rivets may once have held two studs of enamel in settings on the tail[18] which on the Camerton piece (City Museum, Bristol, reg. no. F 684) has become a duck's head, the eye of which consists of a spot and ring of decayed enamel.

Bridle-bits as such have been dealt with by several authorities and only the enamel decoration need be looked at (see Ward Perkins, 1939; Clarke, 1951; Hawkes, 1959; Barker and Megaw, 1963). Only seven of the twenty-six specimens with enamel were not found in the hoards. Those from Hengistbury Head and Llyn Cerrig Bach have a single large setting presumably for enamel[19]. The Rise (Smith, 1925, p. 103, pl. viii, no. 4), Birrenswark[20] and Thames (Read, 1910, pp. 159-160) bits are probably the best known, all with small cells in red and a second colour (only remaining in the blue on the Rise specimen). The pieces in the Saham Toney hoard and also those from the Seven Sisters hoard (see below for details) probably possessed three colours though for the most part these have now decayed or disappeared entirely. The bits in the Polden Hill and Stanwick hoards have red enamel only, either in spots alone or occasionally in small curving cells.

Those pieces of harness equipment generally termed mounts - with two rectangular loops on the back - fall readily into three groups: the quatrelobe, the cruciform[21] and a miscellaneous group. There are seven quatrelobe mounts, from Norton, Suffolk, (Kemble, 1863, p. 195), London (ibid) and the Polden Hill hoard and two each from the Santon and Westhall hoards. The Canterbury mount probably should be included in this group (Leeds, 1933, p. 52, pl. II no. 3). On the whole the main enamel, found in the curving cells, is red relieved by separate spots of blue or yellow. There are six specimens in the cruciform group. There is one from High Rochester (Henry, 1933, p. 103), one from an unknown location now in the British Museum (Smith, 1925, pl. VIII, no. 2) and a third in the Uffizi Museum in Florence (Kemble, 1863, p. 195, pl. XIX, n.5) while one is included in each of the hoards from Middlebie, Saham Toney and Seven Sisters. All have small angular cells with red and a second or even third colour of blue and/or yellow. In addition to these two groups there are various mounts, some with spots only of enamel, found in the Stanwick hoard, others with small cells with two or more colours. There are single finds from Carlisle (Museum and Art Gallery, reg. no. 20.1938), 'Cheltenham'[22], Chepstow (Fox, 1958, p. 129, pl. 70c), and Silchester (Museum and Art Gallery, Reading). Of the four, only that from 'Cheltenham' has red enamel on its own.

Among the Iron Age harness equipment belong some of the button-and-loop fasteners - the majority of which are dealt with under the Roman material. These objects are listed and discussed elsewhere (Wild, 1970, pp. 137-155). Using Wild's categories there are two class I examples from Glastonbury (op.cit, p. 148, no.4) and Silchester (Museum and Art Gallery, Reading) and two 'atypical' specimens rather similar to each other, one from Newstead (op.cit. no.159) and the other from Perth[23]. None of these retains its enamel. One such object is part of the Middlebie hoard and no less than nine are to be found in that from Stanwick.

The classic paper on linch pins is still that of Ward Perkins (1940, pp. 358-367). There are the three well known finds, from Colchester (Hawkes and Hull, 1947, pp. 331-2) with red and blue enamelled scrolls, King's Langley (British Museum, reg. no.1940.2-3.1) with red alone and two from Colne Fen (Tebbutt, 1961, pp. 235-238) - the first in red only, the second in red and blue. Besides these there are two unprovenanced specimens, one in the Ashmolean Museum and the other in the Museum, Saffron Walden (Henry, 1933, p. 88). Six examples of enamelled linch pins are known from the Stanwick Hoard.

There are also nine miscellaneous pieces of harness or chariot equipment which do not fit into any of the previous categories. On most of these the enamel is now missing as from the much published mount with a triskele design now in the Ashmolean Museum (reg. no. N6418) and the somewhat similar object from Dowalton Loch (Agnew, 1881, pp. 154-5). Only on the Cavenham mount (Clarke, 1939, p. 71) is the red enamel preserved as also on that from Fremington Hagg[24] where the blue is obvious but the second enamel has now decayed. The remaining pieces do not call for any comment[25].

Military Equipment

Although not numerically large the military equipment displying enamel constitutes an important part of the Iron Age finds. Most is found on swords or scabbards. The Bugthorpe discs with their rivetted studs of red enamel are generally described as scabbard mountings[26]. The Bugthorpe scabbard itself (Fox, 1958, pp. 41-3, fig. 26) may have had enamel once in the spot and ring on top of each of the rivets. Both the Embleton sword (on the hilt) and scabbard have small cells of enamel in red, blue and yellow though much of this is now decayed[27]. The Thorpe sword hilt has enamel in yellow and blue on the heads of small studs which are applied[28]. From Hod Hill comes another hilt (Brailsford, 1962, p.1, no.A1) with both a simple red inlay and also, but now gone, enamel of two or three colours in tiny cells as on the bands of the Embleton scabbard. A hilt guard also from this site (ibid, no.A2) has red enamel only. The enamel has now gone from the Richborough scabbard terminal (Bushe-Fox, 1949, p. 131) as also from the Little Wittenham chape (Harding, 1972, p. 131) and the Glencotho scabbard (Piggott, 1950, p. 13). The scabbard from Dailly, of which there is a replica in the National Museum of Antiquities of Scotland, has red enamel.

Enamel is found on both examples of Iron Age helmets discovered in Britain. The horned helmet found in the Thames at Waterloo Bridge (now in the British Museum) originally had six - of which five now remain - applied studs with incised heads on some of which are slight traces of red enamel (Fox, 1958, p. 49, pl. 27b x fig. 36a,b). The second helmet, of later date and with a neck guard, is an unprovenanced find, formerly in the Myrick collection, now in the British Museum and considered to have come from the North of England. Originally it had three applied studs on the neck guard and two on the upper part of each ear-flap. These too have incised heads which were once covered with red enamel, traces of which can still be seen on some of the remaining studs - two on the neck and one on each of the ear flaps (ibid. p. 119, pl. 62c).

On the Battersea shield there are twenty-seven applied studs of red enamel. Whether this is in fact an enamel or not and the exact process of applying the studs is discussed later. For the present Megaw's description is the most adequate (Megaw, 1970, pp. 150-1). It is possible that the missing stud, of which the rivet only can be seen, in the centre of the Tal-Y-llyn shield boss was of red enamel (Megaw, 1970, p. 158)

Miscellaneous Types

Moving from the military sphere to more peaceful pursuits we meet four mirrors which have enamel. These are from Birdlip (Fox, 1958, pp. 89 and 99), Old Warden (Spratling, 1970, pp. 9-16), Nijmegen in Holland (Megaw, 1970, pp. 156-7) and Holcombe (Fox and Pollard, 1973, p. 22) and each is inlaid with red enamel[29].

Several more tankard handles seem to have been intended for enamel though little of this now survives. In all cases the enamel occurs as studs either with a rivet through the centre as on the Aylesford specimen (Corcoran, 1952, p. 97) or on the incised domed heads of applied rivets as on one of the handles from the Seven Sisters hoard and just possibly in like manner on the remaining five from the same hoard, the two examples from Hod Hill (Henry, 1933, pp. 81 & 94) and the incomplete specimen from Caerwent (Simpson 1972, p. 330). The Welwyn tankard handle (Hawkes, 1935, pp. 351-354) still retains most of the red enamel studs which appear to have no rivets and to be solid studs instead of covering an applied metal rivet head.

There is too a group of enamelled bucket or bowl mounts. Both the fish-head spout from Felmersham (Watson, 1949, pp. 41 & 50-1) and the two mounts from Leg Piekarski (Megaw, 1963, pp. 27-37) belong to bronze bowls and possess inlays of red enamel. The bull's head mount from Lydney (Wheeler, 1932, pp. 75-6) enamelled in red and blue[30] and the human face mount from Boughton Aluph (Megaw, 1970, p. 133) with its 'fringe' of red enamel would appear to have acted as bucket mounts. The two bucket mounts in the shape of a ram's head from Harpenden (ibid., pp. 161-2) each have two circular settings at the nostrils and sides of the mouths (therefore four on each) for red enamel studs as well as having blue enamel settings for the eyes[31]. The enamel on the Great Chesterford bucket (Henry, 1933, pp. 81-2) however is similar to that on the Waterloo helmet (q.v.) that is to say red on the incised head of an applied stud and with a rivet through the centre. In this group too should be included the Lexden cup (Fox, 1948, pp. 135-6) the handle of which has a single red enamel stud set in a circular setting and held in place by a rivet through the centre and the escutcheon on the handle of the bowl from Snowdon (Savory, 1976, p. 62, no.32, fig.38).

Enamel however is not found very often on pre-Roman fibulae from Britain. There are three fibulae of Iron Age type with this kind of decoration, two being of the involute variety and the third being of La Tene I form. The latter is the well known specimen found at Maiden Castle (Fox, 1958, pp. 158 and 19-23) which possesses three studs of red enamel (one on the foot and one at either end of the spring), once throught to be coral[32]. A red enamel stud with rivet occurs on the involute fibula from Eastburn (Stead, 1965, pp. 47, 65 & 115) while a small portion of red enamel inserted into the foot of the Trevone fibula presents an unusual technique (Dudley and Jope, 1964, pp. 19-23).

Among the Iron Age belt-links only four are enamelled. They are included here though perhaps it is more likely that they constitute military equipment. The incomplete example from Colchester (Hawkes and Hull, 1947, p. 329, no.2) is the odd one out consisting of an openwork figure-of-eight with side loops, the whole having six spots from which the enamel has now gone. The remaining three are more or less similar in shape being rectangular with rounded ends in which there are circular openings presumably for attachment; in front of these are two half bosses and between in the centre is a detachable square enamelled plate. On the incomplete specimen in the Yorkshire Museum (Fox, 1958, p. 119) the plate is now detached. The surface of the plate is covered with rows of interlocked triangles in red and yellow enamel. An example of unknown provenance from the Way collection (Allen, 1904, pp. 169-70) has been similarly treated. The remaining example, from Drumashie (Frazer-Tytler, 1923, pp. 11-13), has four large cells with curving sides, again in red and yellow alternately.

One of the most intriguing of the groups of Iron Age enamelwork consists of the circular hollow domed bronze studs with openwork decoration (in most cases) on the surface filled with red enamel. There are two possible functions for these objects. Two of the hollow and one of the solid types of stud were found with the Lesser Garth terret (q.v.) and in this case appear to fit into the saucer-shaped settings on the edge of the terret[33]. On the other hand the studs found at Hertford Heath (twenty-one in total) and also Lexden (Laver, 1926, pp. 241-254 - four specimens, one still with its backing) would appear to be casket mounts. It is noticeable that the latter are lighter than the former so that it might be possible to use this factor as a criterion for function. At the same time other possibilities for use should be borne in mind e.g. pommels. The specimen from Beckley (Leeds, 1933, p. 44) is a rather crude solid example with four deep spots of red enamel. The twenty-one specimens, all hollow, from Hertford Heath appear to fall into six types though all have only red enamel (British Museum, reg. nos.1958. 7-4. 73 to 93). Compared with these the single find from Hod Hill (British Museum reg. no.92.9-1.528) is a well made hollow stud where however the red enamel holds two circular blue glass spots in place. A somewhat similar stud but with red enamel only was discovered at Ixworth (Ashmolean Museum, reg. no.1927.875). The Lexden studs too are enamelled in red alone and are similar to some of those from Hertford Heath. The solitary stud from Lydney (Wheeler, 1932, p. 75, no.11) is solid with three cells sunk in the surface for red enamel while the example from Wilton Roman villa (Savory, 1966, p. 42) is like the smaller example from Lesser Garth but has now lost most of its red enamel.

Included with the studs are a few of a different type, better known on the Continent (Henry, 1933, p. 79). These are rivets with a domed head which is incised to receive a covering of red enamel. The enamel has now gone from that found at Hengistbury Head (Bushe-Fox, 1915, p. 615) though this may be an actual import. Four such studs, three large, one small, come from Knob's Crook (Fowler, 1965, pp. 36-7). On these the incisions are simple grooves and not cross-hatching and only traces of the red enamel remain. From Hyndford Crannog (Munro 1899, p. 365 - National Museum of Antiquities of Scotland, reg. no. HTA44) comes the shell only of the red enamel with the imprint of cross-hatching visible on the inner surface.

A much better known class of enamelled object is the group of massive bronze armlets from Scotland. In the terminals of these is a circular opening into which was placed a plate of bronze containing enamel. For the most part these enamelled plates have now disappeared and such incomplete armlets are not included here. They are well listed elsewhere (Simpson, 1968, pp. 233-254). However four do survive with the enamel intact. These are a pair each from Castle Newe (Smith, 1881, pp. 330-333) and Drummond Castle[34]. In both cases there is a simple inlay of enamel in two colours, red and yellow. A somewhat similar bracelet but with zoomorphic terminals comes from Altyre[35]; in each terminal is a large circular setting which probably once contained enamel and at each end there are four small spots of blue described as enamel but which could be glass. Another armlet with (?) ram's head terminals each with four settings - two eyes and two nostrils - containing traces of red enamel was found at Snailwell (Fox, 1958, p. 81). A group of four hinged bracelets was unearthed at Mount Batten[36] - each of these contains several pairs of sunken spots in some of which there are traces of red enamel. To finish the bracelets is the gold specimen from Rhayader (Bate, 1866, pp. 502-3 & Archaeologia Cambrensis, 1899, pp. 259-261) with inlays of enamel, now decayed, in curving cells on the terminals.

There is only one certain example of a collar with enamel. This is one of unknown provenance in the British Museum[37] possessing four triangles of inlay on either side of the catch - in these are red and a decayed enamel. These collars as a group have been published (Megaw, 1971, pp. 145-156) and though only this example has sure traces of enamel an open mind should be kept on the original decoration of some of the others.

In addition there are about a dozen various objects which do not fit adequately into any of the preceding categories. Some are mere fragments such as the pieces from Hod Hill (Richmond, 1968, p. 114, no.28) and Knob's Crook (Fowler, 1965, p. 37) both with red enamel, others are somewhat better known. Among the latter is the Elmswell panel (Corder and Hawkes 1940, pp. 338-356) with its narrow bands of red and decayed enamel in curving cells, here however consisting of a vine scroll rather than the more usual Iron Age motifs.

The Hoards

Moving on from the individual finds there are seven hoards of the Iron Age which contain varying amounts of enamelwork[38]. It is only in recent times that these hoards have begun to receive serious attention (Stanwick, Polden Hill, Seven Sisters) but the four lesser finds still await definitive publication. Even in the recent accounts the enamel decoration rarely receives full consideration. Although there are brief mentions of various objects from the hoards in the previous sections it is probably best to treat each hoard separately because of the associations of the specimens in it. Each hoard is dealt with in alphabetical order concentrating on the enamel aspects.

1. Middlebie

There does not appear to be much written on the Middlebie, Dumfriesshire, hoard of horse trappings. Childe (1935, pp. 228, 230, 254) barely mentions it in his text but does provide a plate showing the complete find (ibid, pl. XV). Eleven out of the twenty-seven objects now in the National Museum of Antiquities of Scotland are terrets. Three of the objects are enamelled (reg. nos. 55, 58 and 59 respectively) viz. a cruciform mount with small cells the colours of which are not clear; a terret with four knobs on the edge, each with a spot and ring of blue and red; and a button-and-loop fastener on which the enamel is rather decayed (Wild, op. cit. no. 9).

2. Polden Hill

This hoard was found at the western end of the Polden Hills in Somerset. First noted nearly two hundred years ago (Harford, 1803, pp. 90-93) it has only recently received fuller attention (Brailsford, 1975, pp. 222-234). It was found by chance during ploughing and had seemingly been deposited in a hole in the ground. Over ninety pieces of bronze (and some of iron) consist for the most part of terrets, bridle-bits, cheek-pieces, harness mounts and fibulae. The find is considered to be one of the late Iron Age bronze-founders' hoards. Most of the material is in the British Museum (reg. nos. 46.3-22.64 to 148 and 87.9-6.77 to 79) except for one bridle-bit in the Bristol City Museum. The number after an object mentioned below is the final figure of the registration number.

Altogether there are thirty enamelled objects plus three possible ones. However the definitive account does not discuss the enamel decoration and indeed uses that word only once, an undefined 'inlay' and 'coloured glass' or 'red glass' being used to denote the presence of such material. With one exception these enamelled objects are pieces of horse harness: four 'brooches' or mounts (Fox, 1958, p. 124), five cheek-pieces, ten terrets of group four, ten bridle-bits also lipped and a chape. The only enamel applied is red which survives quite clearly on the harness mounts (78 & 79) and cheek-pieces (138 & 143). In many instances it has now decayed to a creamy-white colour as seen on the jointed harness mount (112/113) and terrets (96 to 98), and in some cases has more or less disappeared entirely e.g. the two bridle-bits (70 & 78-80). Fox suggests (1958, p. 126) the bars of the links of one bridle bit (70) are grooved for enamel but this is unlikely. In a few places, e.g. a cheek-piece (142), the enamel is bubbly but otherwise it seems to have been laid without defect. It is uncertain what the material of the rivetted studs (107) is. Leeds (1933, p. 43) thinks it to be the last use of coral but Piggott (1970, p. 10) seems more correct in thinking it to be chalk though Brailsford (1975, p. 230)notes "? coral". The missing studs of (111) may not therefore have been enamel.

The enamel is a simple inlay in one colour, red. The cells range from spots to much larger trumpet and lobe shapes. Spots are found alone on the harness mount (79) and such terrets as (91) and bridle-bits as (77). They also occur in combination with triangles and scrolls as on (78 & 77). The spots vary from 4 to 7 mms. across and 2 to -1 mms. deep and appear to have been drilled with a centre-bit drill. The enameller was also capable of attaching the enamel to much larger cells as in the flowing patterns on harness brooch (78) or the geometric cells of the jointed harness brooch (112/113).

The Polden Hill hoard provides an excellent series of enamelled bronzes in red alone from spots to scrolls. Whether this is the work of one atelier or a group of objects gathered together over the years is difficult to say. The date of deposition is usually assigned to the middle of the first century A.D.

3. Saham Toney

The best account of the Saham Toney hoard is to be found in the report of its presentation (Grigson, 1849, pp. 398-400) to what is now the Castle Museum, Norwich. Less than a page is devoted to a description of the objects and the best part of the 'article' is the engraving which, however, has been described as doing "hardly justice to the objects" (Victoria County History, 1901, p. 273). The find had been made prior to 1847 and when given to the Norfolk Society by Grigson no details were known.

The discovery yielded seven pieces: a quatrelobe harness mount (no. 1), the ring and shank of a bridle-bit (no. 2) and five turrets (nos. 3-7) all of Leeds' type five (5-7) and six (3 & 4) with knobs on the edge of the ring. All seven objects bear enamel. This has now almost entirely gone from the bridle-bit (2) and on most of the remaining pieces is suffering from decay or corrosion. Red has certainly been used (3 & 4), blue too (1 & 6), yellow (6) and just possibly green as well (3 & 4), though the latter could be one of the other colours which has now decayed to a green hue. The blue enamel found in this hoard is a medium opaque not a light translucent blue (as on the glass insets of e.g. the Hod Hill stud q.v.) and appears to be a true enamel rather than glass. There is no trace of red enamel under it. The cells, with the exception of the knobs on (5, 6 & 7) which are petal-shaped, are mostly all geometric being triangular, rectangular and lozenge-shaped. There is ever only one colour to each cell but on the objects it occurs in the combinations blue and red and red, blue and yellow. The hoard has been associated with the Boudiccan revolt and a date of deposition in the third quarter of the first century A.D. seems likely without trying to be specific about historical events.

4. Santon

The Santon, Norfolk, hoard is best described by Smith (1909, pp. 146-163) though Fox (1923, pp. 104-108) has a useful account of the finds especially the fibulae[39].

Over fifty pieces were discovered in a beaten bronze cauldron by a labourer working in his garden. These consist mostly of bronze objects seemingly belonging to a chariot and to a casket, some fibulae and pieces of harness. It is mostly the latter which provide the small amount of enamelwork in the haord: two quadrilobe harness mounts, half a bridle-bit and a stud (? of a casket-handle).

On the four objects there is only simple red enamel. It is well preserved on the harness mounts. On mount A there are a few spots in addition to the scrolls which rather swamp them while on mount B there are only scrolls and of a more simple nature. On the stud are merely two simple spots. The spots on the two lips of the bridle-bit are empty. Two of these have holes in the centre but this need not mean that rivets held the enamel in place but rather that the centre-bit drill went too deep.

The enamel decoration of the Santon pieces is therefore simple spots and scrolls in red only. The date of deposit, suggested by the fibulae, is considered to be somewhere between A.D. 40 and 60. The objects are housed in the University Museum of Archaeology and Ethnology, Cambridge.

5. Seven Sisters

The Seven Sisters or Neath (parish of Upper Dulais, Glamorganshire) hoard has at last received definitive publication (Davis and Spratling, 1976, pp. 126-147) though the account of its discovery and description is better than most (Allen, 1905, pp. 127-145). It is housed in the National Museum of Wales now under the reg. nos. 04-127 ff. and numbers in brackets after an object here refer to this final number[40].

Although found in 1875 the hoard was not recognised as such until 1902. It was retrieved (in toto?) from a stream and the finders, farm children, may have lost some pieces before their discovery came under academic scrutiny. Thirty-five objects are known, mostly pieces of horse harness, tankard handles and a smith's trade marks - bronze ingots, billets, casting jets, weight and scrap pieces.

Nine of the objects had been enamelled; two terrets (127 & 128), two side links of a bridle-bit (125 & 126), two harness mounts (131 and the British Museum specimen), 2 V-shaped mounts (137 & 136) and a tankard handle (141 - it is difficult to know whether the other five tankard handles were ever enamelled). Except on (137 & 138) where it is reasonably well preserved the enamel is now decayed or gone. Although traces of red enamel are said to have been visible on the centre stud of the handle (141) (Corcoran, 1952, p. 100) there is no longer any trace of it. The red enamel - used alone - on the terminals of the trace hooks (136 & 137) is the usual opaque red. The red on the other five objects is quite obvious but the second colour (or colours) is too decayed to be certain what it was; it could have been white, blue or yellow in the petals of the knobs of the two terrets (127 & 128) and though blue is possible, Allen (1905, p. 129) suggests white and on the modern reproduction in the National Museum of Wales it is shown as white also (Davis and Spratling, 1976, p. 131 give white). This could be another case of blue turning white as on the Westhall mounts though the present state of the 'white' on the Seven Sisters pieces is not the ivory white of the Westhall enamel. The second colour of the bridle-bitand harness mount (125, 126, 131) is given as yellow by Leeds but for (131) at least as bluish white by Davis and Spratling (1975, p. 131). However on (125) some of the enamel seems blue though there is no reasonwhy red, yellow and blue should not occur together as on the Saham Toney pieces. It is difficult to be sure of some of the second colours and Leeds' reconstructions of pieces (125 & 131) (Leeds, 1933, pl. II nos. 4 & 5) should be treated with caution.

The techniques used are quite varied. On the tankard handle (141) the enamel occurs on the incised head of the stud which is set into the circular setting prepared for it in the casting with its rivet through the rivet-hole in the centre. Five of these were once on the handle. The settings in the remaining handles are different and are hardly intended for enamelled studs. On the trace hooks (137 & 138) we find the use of a single colour - red - in spots combined with curved cells while on the remaining pieces there are simple geometric cells with enamels in two or three colours probably red, blue and yellow.

It would seem that here is yet another bronze-worker's hoard which is likely to have been deposited between A.D. 50-70.

6. Stanwick

The Stanwick hoard is the largest of the Iron Age metalwork hoards containing enamel (MacGregor, 1962). It is in the British Museum (with a few exceptions) under the registration numbers 47.2-8 - and 96.1-20 -: the number in brackets after an object mentioned here completes the registration number but only (3) & (4) belong to the latter group.

The hoard was discovered about 1844 seemingly outside the famous fortifications after which it is named. Over one hundred and fifty pieces of bronze, mostly horse harness equipment, plus some iron and non-metallic objects were recovered. Many of these are faultily cast, distorted or broken suggesting a bronze-founder's hoard. Fifty-three of the bronze objects retain or have provision for enamel. Where this remains it is generally a decayed bubbly substance of a greenish colour (? also due to faulty manufacture) but enough remains to show that it was all once red. In the majority of cases the enamel is contained in circular spots varying between 1 and 6 mms. in diameter and ranging in number from one to fifty-two. There are three examples of circular settings with rivets or rivet-holes though in one case (34) where the enamel remains in such settings there is no trace of a rivet head. There is also one example of a spot and circle of enamel (4) and one case of the cells being rectangles (fragment - no reg. no.). The spots seem to have been drilled in the model before casting, since on one faulty piece (75) they are already present. The enamelled pieces consist of terrets, bridle-bits, linch pins, harness mounts and miscellaneous objects.

The hoard is said to comprise four sets of horse harness numbered A-D and buried possibly as late as A.D. 70 (MacGregor, op.cit. pp. 20-1 & 36) though mid-first century A.D. is generally offered as the date of deposition. While the find is adequately discussed by MacGregor (ibid) little mention is made of the enamel decoration. Though in truth not a great deal can be said about empty spots. These however hardly justify the description of the set D as "the controlled use of enamel in comparatively large expanses". In the catalogue (ibid. pp. 37-51) enamel is barely mentioned especially on page 38 where only one of fourteen enamelled objects is actually stated to contain it though the illustrations mitigate this.

7. The Westhall Hoard

The Westhall hoard was first exhibited to the Society of Antiquaries in 1854 (Harrod, 1855, pp. 454-456) and though listed then was better dealt with nearly a century later (Clarke, 1939, pp. 68-9). However two pages cannot adequately cover such a find. This is in the British Museum,reg. nos. 1855.5-19.1-35,and again reference is made to the final part of this official number.

The objects were found by accident during drainage operations which revealed a habitation area covering about two acres. The hoard seems to have been contained in a bronze vessel. The association with the coin of Faustina the Elder and the sherd of Samian ware is no longer accepted. Besides the enamelled objects there are six plain linch pin heads, a lamp dated to the third quarter of the first century A.D. (Clarke, 1939, p. 69), various fragments of bronze and iron, three polished flint pebbles and a disc brooch (not closely datable). There are ten enamelled objects consisting of eight terrets of Leeds' class 3 (broad faces) and two quatrelobe harness mounts. While these can be generally said to belong to the mid-first century A.D. the deposition of the hoard is normally associated with the Boudiccan revolt.

On the terrets the enamel is all red and quite simple. A lot of it has disappeared or is decayed now but it was obviously all once red. The scrolled cells on (1-5) are quite large but cover an even larger area on (6-8) where the bronze palmettes take up only the centre of the broad field leaving large areas for enamel on either side - particularly see (6). Even in the remaining bronze parts there are spots of enamel, again apparently red.

The two quatrelobe harness mounts are interesting (9 & 10). The main cells, 'three-sided voids', contain red enamel but there are also seventeen spots. The colour of these is even more interesting. At present they present three shades viz. blue, blue tinged with various amounts of white and white alone. The blue is medium and translucent while the white is opaque and ivory-like. The spots were therefore generally described as being blue and white (Kemble, 1863, p. 195) until it was pointed out that some of the white pieces contained blue and that the white was in fact once blue which had become devitrified (Leeds, 1933, p. 41). There are however occasional relapses (Megaw, 1970, p. 163). Leeds' explanation is the most logical yet the white is quite unlike any normally decayed enamel which on turning white becomes dull and off-white tending towards cream or green. Here it is quite ivory-like but this may be explained by the fact that the spots are decayed glass insets not true enamels. That is, the glass spots are held in place by red enamel as on the Suffolk terret (q.v.) or Hod Hill studs (q.v.). On the Westhall pieces the glass insets are a good fit and no red is visible round their edges but on analogy with other examples of translucent blue spots this would appear to be the answer.

With this examination of the Westhall hoard the survey of the pre-Roman enamelwork from Britain is complete.

SURVEY OF ROMANO-BRITISH ENAMELWORK

INTRODUCTION

The following three chapters deal with those Romano-British objects which are enamelled or appear to have been intended to hold enamel decoration. Overall approximately eighteen hundred were noted and are now discussed.

The pieces are dealt with by type. Chapter 2 covers bow brooches, Chapter 3 plate brooches and Chapter 4 non-brooch material. Within these headings each individual type of object is described with particular reference to the enamel on it - where this occurs, what colours are used, what designs the cells take and how the enamel is attached; brief comments are added on dated examples and distribution patterns.

Yet is is unlikely that all the enamelwork found in Britain was made here. This is not the place to attempt a survey of Continental enamelwork but some brief reference does need to be made in order to distinguish between those enamelled objects which were made in Britain and those imported from the enamel-working centres of Gaul and the Rhineland. The best general survey of the subject remains that of Henry (1933, pp. 66-146) who discusses all types of objects from western Europe and an idea of the actual material may be gained from the diverse but comprehensive collection of enamelled objects housed in the Ashmolean Museum. The products of the Rhineland are well known from Exner's work on the brooches (Exner, 1938, pp. 31-121) and Sellye's on both the brooch and non-brooch material from Panonia, generally considered to have been imported from the Rhineland (Sellye, 1939, pp. 1-91). Large quantities of enamelled objects have been recovered, especially during the nineteenth century, in the Namur region of Belgic Gaul and form the basis of Bequet's study of the 'jewellery' of the area (Bequet, 1937, pp. 261-274); the collection housed in the National Museum in Brussels has also been described (De Loe, 1937, pp. 261-271). From further south, from the territory of the Sequani, comes the group of brooches now in the museum at Besancon (Lerat, 1956) and among the brooches from individual sites are those recovered at Augst (Rhia, 1979). From somewhat further afield comes the small but interesting group of enamelwork discovered at Dura Europos (Toll, 1949, pp. 35-43).

An examination of the brooches discussed by Collingwood on the one hand and Exner and Sellye on the other is enough to show two quite different sets of objects. The Rhenish enamelled brooches are generally more geometric and flamboyant than their British counterparts; the former are square, rectangular, lozenge-shaped, rhomboidal, wheel-shaped and so on; they are frequently "stepped"; and often possess additions in the form of debased zoomorphic heads (especially over the foot and as arms), knobs or lugs around the edge or some sort of stud standing upright in the centre (often missing now). Again the bow and plate brooches from Namur are not types generally met with from Britain except in very small numbers e.g. those with a flat triangular bow and narrow zoomorphic foot; the plates are mostly lozenge-shaped with projections at the angles or large open discs, best described as wheel-brooches. However there are plainer disc brooches and animal plate brooches from Gaul which are more like the British finds. These as Wheeler has remarked form part of a widely produced Roman group equally at home in Britain or Belgic Gaul.

The problem of a wide source of origin applies more especially to the non-brooch material. The seal box lids, the miscellany of mounts and studs, the belt-plates were all made somewhere but was it in Britain, Gaul or the Rhineland? The suggestion is put forward that some of these were made in Britain and that all those found here do not represent the flooding of the new province with enamelled objects from other parts of the Empire. It is however something about which we cannot be dogmatic for there are few obvious differences in the same way as there are between the brooches. Certain groups however appear not to be British such as the rectangular mounts with a knob at the top, the vessels with a polygonal design, some types of seal box lid, probably the open rectangular belt-plates and a few individual pieces. What is clear however is that (a) there is enough material which can be satisfactorily regarded as Romano-British in origin and which can be used to say something more about the craft; and (b) the imports are useful in order to make certain deductions about the possible influence these pieces had on the native workers.

CHAPTER 2

ENAMELLED BOW BROOCHES

This chapter covers the most common and perhaps the most British type of enamelled object. There are three main groups of bow - as opposed to plate - brooches: the head stud, group M and the trumpet brooch, sub-divided into groups R and S. The majority of the approximately six hundred enamelled bow brooches examined belong to these groups, each roughly equal in size. Besides these there are small numbers of various other groups which on the whole do not bear enamel decoration. These include head stud varients, dolphin, knee, thistle and other miscellaneous types. For the most part Collingwood's arrangement of these brooches is followed since it has generally stood the test of time and has so far not been replaced by anything better[41]. It should be remembered that only enamelled specimens are discussed here and that remarks made may not apply to a group as a whole.

Group Q - the head stud brooch (Figure 2A)

The head stud brooch constitutes Collingwood's group Q, also sometimes known as the Lamberton Moor type. The group is among the commonest of Romano-British brooches and nearly one hundred and fifty enamelled specimens have been noted. Out of fifty-four Q brooches cited elsewhere about three quarters bear enamel decoration and of the thirteen unenamelled specimens seven are from Traprain Law (Hildyard, 1954, p. 73).

The features are given in the classic account of the type (Collingwood and Richmond, 1969, p. 296, nos. 39-45). The brooch is described as being stout and heavy, with an almost semi-circular bow, the foot of which expands into a moulded knob with flat face. At the head is a stud derived from a rivet which originally helped hold the spring in position; the head loop and shackle, of loose wire or cast solid, is a common feature. The type is thought to be of northern manufacture with southern examples being imitations. Early examples are now thought to have first appeared during the third quarter of the first century and the type continued to be manufactured well into the second. Recently a detailed study has been made of the twenty-one group Q brooches in the British Museum and in some cases the enamels were analysed (Painter and Sax, 1970, pp. 153-174). On the whole the results tend to confirm Collingwood's comments on the technical features but it is held to be too small a group - and too varied - to confirm his other assumptions (*ibid.* p. 173). The classic example of the type may be taken to be the pair found in a hoard on Lamberton Moor, Berwickshire[42].

In all over one hundred and fifty head stud brooches have been encountered which either are still enamelled or once contained some form of enamel decoration on them (figure 2A). The most common place for the enamel is in a strip running along the bow from the head stud to the foot. About one fifth of the specimens have the arms enamelled and on a mere half dozen the spring covers or collars are likewise treated. The head studs on approximately two thirds of the group appear to have been enamelled. About thirty examples have had enamel in the face of the foot instead of this being plain. In some of the last two instances all that is left is a mere cast cavity at the head or in the foot knob with a rivet hole obviously intended to hold a rivet, the head of which may have been enamelled in the same way as on cast examples. However there is the specimen from Richborough where the head and foot studs are described as being enamelled in yellow with four dark spots (Bushe-Fox, 1949, p. 114, no. 34, pl. XXVIII). The decoration is in fact not enamel at all but a 'bead' of bone or some chalky (?) substance, now yellow, set into the two cavities and held in place by a bronze rivet, the bronze head of which is dished and has four bronze spots! Thus it cannot be certain that a cavity implies the use of enamel. Just over thirty examples have plain bows with the enamel confined to the studs or arms - usually both studs but where only one, overwhelmingly the head stud. Some specimens such as one from Richborough have enamel in all these places - bow, arms, spring cover, head stud and face of foot knob (Bushe-Fox, 1949, p. 114, no. 35).

Blue and red are the most common colours, surviving in almost equal numbers[43]. Yellow occurs less than a dozen times and orange, if so intended, only once. White is found on half a dozen examples and on some at least appears to be an intentional colour though in others could be the result of decay. Likewise green is found on a few specimens but here the intention is more doubtful[44]. On about half the Q brooches examined the enamel has either decayed or disappeared entirely - roughly equal numbers in each case. Where two colours survive together the most frequent combination is red and blue and less frequently blue and yellow with blue the dominant colour in each case.

The most common design on the bow strip is a lattice pattern made up of a row of lozenges or diamonds along the centre with triangles down each side. The number, and hence the size of the cells, varies greatly from (for the lozenges) two up to eighteen with some examples of almost every number in between but with around a dozen perhaps slightly more usual. As a general rule the number of triangles is double that of the lozenges plus two. The main colour appears to be blue for the lozenges and red for the triangles. The next most frequent design is a row of squares or rectangles (the two cannot always be readily distinguished) ranging in number from five to fifteen. Whereas the lattice pattern occurs on over a third of the specimens noted the latter is found on slightly less than half that number. On a dozen specimens the centre of the bow is taken up by a row of bronze lozenges thus leaving either a row of triangles on each side for enamel or else two long cells with the upper edge of each 'toothed'. It is interesting to note that on this type a ridge often runs along the crest of the bow (i.e. along the row of lozenges) and that sometimes in the type where the lozenges are cells (the lattice design) the enamel there is so ridged too though there is no technical necessity for its being so. The arms where enamelled usually have simple cells involving lozenges and triangles for the most part. It is similar on the spring covers. The enamel on the head studs and likewise on the faces of the foot knobs is contained in a simple spot, a spot and circle or merely a circle - less frequently there is a more elaborate design on the head stud.

The enamel on the vast majority of head stud brooches is found as one colour to a cell though generally two colours occur on the one specimen. There is no example with more than two colours. On less than a dozen of the total a more complex enamel is found, generally on the bow where the strip is one long cell filled with squarish pieces of enamel alternately red and blue with usually about ten pieces in all without any metal separating the colours. This is well seen on two specimens from Newstead (Curle, 1911, pp. 323-324, pl. LXXXVI, nos. 21 and 22) though on a couple of other examples the enamel on the arms consists of a ground or bed of one colour with a few spots of a second. No millefiori enamel was met with.

Since the majority of Q brooches seem to have had some sort of enamel applied we are probably more justified in this instance in looking at Collingwood's statement that they are mostly of northern manufacture. In fact only a third come from sites in the north though this in itself may be a large enough number to claim manufacture there and of course the distinction must be made between the area of manufacture and that of deposition. Yet about two thirds have been found in what might be termed the south with a fairly even spread throughout this area. The area of manufacture is therefore perhaps more widely spread than Collingwood suggests though it would need a much deeper analysis of the find-spots and objects to be certain.

Likewise the earliest of these brooches must now be put further back than Collingwood suggests. Three have been found at the Lunt in a pit dated to before A.D. 75 (Hobley, 1973, p. 66) and it might be added that one of these has the more complex enamel mentioned above. Another from Richborough was also found in a pit, the date of which is placed between A.D. 75 and 90 (Bushe-Fox, 1949, p. 114, no. 35) though elsewhere on the site a specimen was recovered from what appears to be a very late third century context (ibid, no. 38) and the Old Winteringham example also seems to have been lost during the course of that century (Stead, 1976, p. 198, no. 12). However date of manufacture and date of deposition are two different things which may not closely coincide. Between these extremes are about a further score of dated examples belonging mostly to the second century such as the Antonine ones from Newstead, possibly those found in the lower levels of Traprain Law or the specimens from the Walbrook dropped before A.D. 155 (Merriefield, 1962, pp. 36-52).

Also to be noticed is the variety of sites from which this type of brooch has come, fort, metropolis and native site.

Q. Varient

A variety of brooch which stands out is probably related however to the Q group. Its main features which place it apart are a row of small knobs or spikes along each edge of the middle of the bow and a prominent crest, often in the shape of a recumbent dog (head as head stud). There is a good example of the type from the Lunt (Hobley, 1973, p. 65). Hull calls this the 'sawfish type'. Twelve examples have been noted. Most have enamel on the bow where, like the main Q group, the design of lozenges and triangles prevails. Much of the enamel is decayed or has disappeared but red and yellow seem certain.

All of it has been applied simply as one colour to a cell. No specimen has been found in the north and in fact most come from the south east of the country. Several are dated. Two from the Lunt are dated earlier than A.D. 75 and a parallel from Broxtowe, "an equally early site", is known (Hobley, op.cit. p. 65, fig. 19,1 and 4). One from St. Albans comes from a very early second century deposit (Frere, 1972, p. 116, no. 12) and another from Wroxeter appears to be Flavian (Atkinson, 1942, p. 203). A second century specimen is recorded from the Walbrook.

Group M (Figure 2B)

Even more numerous than the head stud variety are those brooches of Collingwood's group M, to which no better term has been applied. Just over two hundred enamelled specimens were examined and many plain examples were met with.

Collingwood (Collingwood and Richmond, 1969, pp. 295-296 nos. 26-31) has little to say about his group M though in truth they are very simple. The main feature is the tapering bow starting rather wide at the head and narrowing towards the foot. It can be either hinged or sprung, frequently has a head loop, and possesses either a bent or angular bow generally broken in the middle by a sudden waisting, a simple pair of wings or a poor attempt at an acanthus. He adds little on the enamel which could be summed up as 'crude'. Hull when dealing with the Nornour examples ignores the M typology but does not add anything better in its place (Hull, 1967, pp. 32-35).

A detailed study of the type, enamelled and plain (of which there are more than Collingwood seems to suggest) is necessary but inappropriate here. The former can be classified into five groups depending on whether the bow is bent or angular (figure 2B). Subgroup (a) has a reasonably plain bent bow sometimes with a raised (purely decorative) head stud as with Collingwood's no. 26 (figure 2Ba); (b) a bent bow with a sudden waisting in the middle as no. 28 (ibid.) (figure 2Bb); (c) a bent bow which expands to a roughly lozenge-shaped centre (not specifically mentioned by Collingwood unless his winged type but see Hull's Nornour type 12) (figure 2Bc); (d) an angular bow with crude decoration on the wider upper end of the bow as Collingwood's no. 27 (figure 2Bd); (e) an angular bow where the upper end contains one long rectangular cell (ibid. no. 29) (figure 2Be).

The enamel on the group M brooches is nearly always found on the upper bow and nowhere else except in the centre of the bow on the subgroup (c) type. Where there is a raised head stud, as on subgroup (a), this is often enamelled. The cells into which the enamel is placed are of the simplest kind, of small size and generally shallow - the last reason probably accounting for the great loss of enamel from the whole type. The most frequent cell shapes are triangles and lozenges, sometimes alone but usually in combination especially on subgroups (b) and (c); small squares and rectangles also occur. On type (e) the upper bow is divided into lozenges and triangles or else into two rectangles running longitudinally and separated by a septum. There are odd exceptions to the normal designs. Occasionally the only enamel is in a lozenge-shaped or oval cell - if it can be thus called since it is so small - on the raised head stud but generally it is sunk into the surface of the bow.

On well over half the specimens examined the enamel has decayed or is entirely lost. The survival rate is poor due probably to the lack of anything to hold on to. Red is the most frequent of the colours still in evidence followed closely by blue and at some remove by yellow and/or orange (since it is not always clear which of the two it is). Green is present on half a dozen examples though in only one or two cases does it seem to be intentional rather than a decayed remnant of another colour and the very few instances of white apparently also constitute a decayed colour. Blue and red is the combination where rarely two colours survive on the same object though instances of either of these with orange/yellow have been noticed. It would appear that only three colours were certainly applied with possibly green in a few cases. All three occur on examples of each subgroup.

With the exception of group (e) the enamel is always one colour to a cell (often the cell) but there are also different coloured enamels on the same object in separate cells. The enamel decoration on the group M brooches is very simple. Only on group (e), and then on half the noted examples, is there a more complex technique with up to five tiny pieces of two differently coloured enamels sharing the one long narrow rectangle without any metal septa keeping them apart.

Very few of these brooches can be closely dated. The two from the Caerleon amphitheatre are dated c. A.D. 125 and c. A.D. 130-160 (Wheeler, 1928, p. 182). Those from the baths at Caerwent are said to be Antonine (Nash-Williams, 1930, p. 240) while one from Wroxeter is dated to the first half of the second century. Another example from Shakenoak came from a deposit dated to A.D. 100+ - 150 and a second was discovered in a level assigned to somewhere between A.D. 240 and 370 though was considered, probably rightly, to be a stray from the second century (Brodribb et al. 1968, pp. 82, 94-5). The three from the Chew Valley Lake excavations were found in contexts assigned to the late first to second, the second and late third to mid-fourth centuries respectively (Rhatz and Greenfield, 1977, pp. 293-299, nos. 8, 14 and 10). Thus most are assignable to the first half of the second century A.D. There is as yet none which can be said to be first-century though the number of dated examples is so small that this should not be ruled out. The majority of the M brooches which come from Charterhouse and Nornour are undated - certainly those from the former site although those from the latter are assigned by Hull to the late first to early second century.

The distribution of the type is interesting. They are very obviously centred in the south, and especially north Somerset, where Charterhouse alone has yielded sixty-three examples (mostly in the City Museum, Bristol). They spread into Oxfordshire where Woodeaton has yielded eleven (Taylor, 1917, pp. 107-111, Kirk, 1949, p. 11), as well as across the Bristol Channel to Caerwent and Caerleon which together account for another dozen specimens. Outside this area forty-one come from Nornour alone and seven more from Silchester. There are a few strays from the Midlands and even less from further north. Very few come from military sites, few from town sites, religious sites such as Woodeaton provide some[45] and Charterhouse so many because possibly they were manufactured there.

Overall group M seems to represent brooches enamelled at a somewhat low level of ability in the south-west during the first half of the second century. It is a reminder of the pieces from the Polden Hill and Seven Sisters hoards and possibly represents the final stages of enamel-working in the area.

Trumpet brooches - group R (Figure 3A)

Not quite so numerous as the last two groups is the trumpet head variety, designated the letter R by Collingwood (Collingwood and Richmond, 1969, pp. 296-297, nos. 46-60). This is sub-divided into four groups according to the shape of the moulding in the centre of the bow: (i) has a plain knob the whole way round; (ii) has an acanthus the whole way round; (iii) has only half a plain knob on the upper half; (iv) has only half an acanthus on the upper half. There is a variety of the last (op.cit. no. 60) with a large plate between the head (now narrow rather than a trumpet) and head loop. Since this group is important from the point of view of the enamel it is termed (iva) here. Otherwise the standard classification is followed.

There are just over eighty enamelled specimens of group R brooches. There appear to be no examples of subgroup (iii) and only a few questionable examples of subgroup (i) bearing this form of decoration[46]. Subgroups (ii) and (iv) are however frequently enamelled (figure 3A) in roughly equal numbers though what proportion these are in to plain members of the group is unknown. The special subgroup (iva) always seems to bear enamel. The decoration is usually found on the head and lower bow together on the one specimen. Seven examples have been noted with the spring cover so treated and there is one occasion each when enamel is to be seen on the catch-plate and on the outside of the foot knob. About half the brooches examined have plain faces to the foot knob. The remaining half appear to have or have had enamel here and in about half of these only here. Again in many cases the foot knob is a cavity with the stud/rivet now missing and it is not always safe to assume that the head of this bore enamel since one example of a R(i) brooch from Wroxeter has a conelian set in the foot (Bushe-Fox, 1913, p. 26, no. 6).

The enamel has decayed or disappeared from less than half the specimens - a much better survival rate than on the group M brooches and suggestive of better workmanship. Where it survives it is generally as a quite recognisable colour. Red and blue are the most common of these but yellow (sometimes an orange-yellow but not orange enough to justify calling it so) is not far behind. The three frequently appear together especially on the trumpet head. There are however some more certain cases of orange rather than yellow on the varient group (iva). There is one alleged example of black on a brooch from Wilderspool (May, 1904, p. 81). The white on one from Silchester (Cotton, 1947, p. 145, fig. 8,3) may be a decayed enamel as the blue, now almost white, on one from Great Chesterford[47].

The designs of the cells are relatively simple. On the trumpet heads most of the upper surface is enamelled either in the more frequent geometric cells found on the R(iv) brooches for the most part or a more Celtic-like curving ornament which occurs less often and only on the R(ii) variety as on that from Spynie (Robertson, 1970, p. 218, 10/4). The most frequent geometric design is where the head is divided into two down the middle and in the ground on each side is a further triangular and semi-circular cell. Usually the ground is red, the triangles yellow and the semi-circles blue as on one from Traprain Law (Cree and Curle, 1922, p. 250, fig. 28, no. 2). Other geometric cells consist of small spots and ovals. On both subgroups the basic cell on the bow is the triangle usually in a row down each side but often a row of interlocked triangles. Sometimes the work is very poor and there is really only one long cell each side of the bow with the upper edge zig-zig or toothed in each case. Very rarely are there a few squares instead. The design on the face of the foot is always a spot, a ring or a spot and ring. On subgroup (iva) the head has a long narrow rectangle divided longitudinally to give two or three cells. On the whole group R cells are deeper than those met with before and this may help explain the better survival of the enamel on this type.

On the specimens of R(ii) and (iv) brooches the enamel is always of the simple one colour to a cell type though usually with three colours used on the one object. Neither complex nor millefiori enamels occur. However subgroup (iva) stands apart in that it always seems to possess - where it survives - a more complex variety of two colours (usually orange and either red or blue) used together in each long narrow rectangular cell on the head. The pieces of enamel used are exceedingly small but the edges between them are quite clear and there is no trace of a metal divide. On an example from Wilderspool even the two major septa are missing thus giving three rows of eight pieces alternately orange and a decayed colour in the one cell (Thompson, 1965, p. 83)[48]. This use of a complex enamel in long rectangular cells on the head is so reminiscent of that on subgroup (e) of Group M (q.v.) that there must be some connection between these two varieties. Thus for the trumpet brooch as a whole simple enamels are the norm with this exception of a complex technique used on the (iva) varient - and there is a complete absence of any millefiori decoration.

The dated examples are few. Newstead has provided some reasonably dated specimens of R(ii). Two appear to be from the first century occupation of the site (Curle, 1911, p. 322, pl. LXXXV, nos. 11 and 12) and two seem to be Antonine (ibid. nos. 13 and 14). The baths at Mumrills yielded another associated with early Antonine pottery (MacDonald and Curle, 1929, pp. 469, 554-5, fig. 115,1). One from the Lunt can hardly have been deposited after A.D. 75 (Hobley, 1967, pp. 110-111). The Lydney Park specimen is assigned to the period A.D. 90-150 (Wheeler, 1932, pp. 76-7, fig. 12,17). The remaining examples are not from dated contexts. However it would appear that the R(ii) type was being made perhaps as early as A.D. 50 and was in use to the mid-second century at least[49]. Most of the R(iv) specimens belong to the early decades of the second century. One from the Caerleon amphitheatre was found with pottery dated c. A.D. 100 (Wheeler, 1928, pp. 162-3, no. 14) while an example from Richborough is said to be Antonine (Cunliffe, 1968, p. 82, no. 31). Silchester has yielded one dated A.D. 120-170 (Cotton, 1947, p. 145, fig. 8,2). The stratified specimen from Wroxeter was recovered from a filling placed between A.D. 100 and 130 (Atkinson, 1942, p. 206, no. H65). Again the remainder are largely from contexts of which the dates are unknown or poorly worked out. Only two of the subgroup R(iva) come from dated layers, that is the specimen from Silchester from a level open between A.D. 100 and 120 (Cotton, 1947, p. 145, fig. 8,3) and the incomplete example lost at Biglands milefortlet in the 150s (Potter,

1977, p. 171, no. 16). Possibly therefore R(ii) was in use before R(iv) especially in view of the Celtic-type designs on some of the former as opposed to the geometric patterns of the latter.

The distribution of the enamelled R brooches calls for note. Over eighty per cent were found south of a line from the Humber to Formby Head. Of those from the north of this line half come from Newstead and Traprain Law. This is rather at varience with Collingwood's statement (op.cit. p. 297) that they are particularly common in the north with their numbers diminishing towards the south. However there appears to be evidence for northern manufacture and this is dealt with later. Subgroup (iva) stands out as having a wholly southern distribution and this combined with the complex enamel similar to that on the M subgroup (e) might suggest it is a product of the south-west though only one has been found at Charterhouse and none at Nornour.

Group S (Figure 3B)

The trumpet head brooch lacking the acanthus moulding or its equivalent is given the letter S by Collingwood (Collingwood and Richmond, 1969, pp. 297-8, nos. 61-8)[50]. This is subdivided into three groups: (i) with a moulding in the centre and a roughly fan-tailed shaped foot (61-3); (ii) with a flat plate as the centre of the brooch - this plate can be circular as (64), semi-circular as (65), pelta-shaped (66) and in the form of wings (67)[51]; (iii) has a circular plate in the centre of the bow and a fan-tail shaped foot (68) though here the head is not always trumpet-shaped. The only work on the group since Collingwood's is that on the subgroup S(ii) type as (64) i.e. with a circular plate and trumpet head (Richardson, 1960, pp. 200-213). One specimen of subgroup S(iii) from Kilpedher is important and is dealt with on its own.

There are approximately eighty enamelled S brooches. Subgroup (i) is rarely enamelled and only five examples were noted. In each case the decoration is to be found on the fan-tail (figure 3Ba). The S(ii) subgroup is the most common one with enamel. The types with the circular plate (64) (figure 3Bb) and the 'wings' (67) (figure 3Be) respectively possess it far more frequently than the varieties with the pelta (65) (figure 3Bd) and semi-circular plate (66) (figure 3Bc) which between them muster less than a dozen examples (even then it is questionable if some of the decoration is enamel). In all four varieties of subgroup S(ii) the enamel is found on the plate on the bow. There are only two occasions when there may have been enamel at the end of the foot such as on the one from Richborough "set with an enamel stud" (Bushe-Fox, 1949, p. 115, no. 40) though this is difficult to be sure about. On the S(iii) subgroup (figure 3Bf) enamel is found on over a score of examples, on most on both the plate and the fan-tail but in a few cases on only one or the other. There are a few instances of enamel on the arms and only one on which the head bears enamel. However on the S group as a whole the trumpet head, unlike that of the group R brooches, never has enamel applied to it. What decoration there is on the head is generally niello or silver inlay which has sometimes been confused with enamel (e.g. Burley, 1955, p. 160, no. 46).

For the group as a whole blue and then red are by far the most frequent colours. There are a few examples of yellow, white, orange and green. The enamel found on the few S(i) specimens is red and blue with a decayed colour, probably originally either of these. That on subgroup S(ii) is quite well preserved - only a quarter present decayed or absent colours. Where it is decayed the enamel is frequently a green colour but in a few cases the green does seem to be the original colour. Over twice as much blue is still present as there is red and yellow occurs only twice. The red and blue are generally found together. In some cases the blue is a rather weak colour and has almost turned white due perhaps to decay as much as to poor colouring technique. On subgroup S(iii) a greater proportion of the enamel has decayed or gone completely but enough survives to show that red, blue and yellow were applied and possibly green too. Again a combination of two of these colours is the norm.

For the most part the designs of the enamelled cells on the S brooches are quite simple. The circular plates on S(ii) (64) and S(iii) (68) have either just a ring of enamel or a spot and one or two rings. The usual pattern on S(iii) is a spot and one ring with triangles and/or lozenges common in the fan-tail. On subgroup (ii) (66) the pelta-shaped plate is divided into two cells while on variety (67) each 'wing' constitutes a cell, normally filled with blue but red and orange were each noted once and often there is a third cell between them. While there are some exceptions there is nothing out of the ordinary though the tendency for some cells, e.g. the 'wings', to be larger than those previously met with might be noted.

In the great majority of these brooches the enamel is one colour to a simple cell though usually with two colours to the object except in the case of the subgroup (ii) varieties (66 & 67) where the colour is the same in both cells. There are however three specimens with a more complex enamel. The first is on a S(i) brooch which appears to have one narrow rectangle on the fan-tail containing five small pieces of two types of enamel without metal divides (= Collingwood's actual example no. 63). Two examples of S(ii) also have a complex enamel: one from Wroxeter as variety (64) has a large spot in the centre plate containing eight pieces of enamel now decayed but obviously once two different colours (Kenyon, 1938, p. 224, no. 3) while the other from Pentre (Davies, 1949, p. 126, no. 13) has on its semi-circular plate (as 65) a ground of one colour with three spots of another added to it. In addition there is one case of millefiori decoration on a subgroup (iii) brooch from Kilpedher.

The Kilpedher brooch is both exceptional and important (Lethbridge, 1952, pp. 182-184, fig. 1,4). According to the excavation report the pelta-shaped fan-tail contained a slice of millefiori glass[52], consisting of rows of triangles alternately yellow and black, set in a red ground i.e. a bed of red enamel was laid first and then the millefiori glass was set on top and the two fused. Unfortunately, we are told, the millefiori fell away just after the brooch had been drawn. Examination of the brooch (National Museum of Antiquities of Scotland, reg. no. GS84 - 1958.615) shows the fan-tail cell filled with a decayed substance, the surface of which is pitted with triangular hollows - presumably the impressions made from the millefiori which has not survived. However the important thing is that here is a British type of brooch bearing millefiori enamel (i.e. where the millefiori glass floats on a layer of red enamel) and found on South Uist. It would seem reasonable to infer that this is a British product and that such enamelwork was produced in Britain. The object is dated to the second century[53].

Dated examples of the group are most common from subgroup (ii). Only one S(i) specimen, from Lydney Park, has been given a date and then only tentatively in the third century (Wheeler, 1932, p. 77, no. 20). There is no need to repeat the evidence relating to subgroup S(ii) variety (64): it appears to have been in use from before A.D. 150 up to the end of that century (Richardson, 1960, p. 206). The only datable example of (65) comes from the Jewry Wall site in Leicester and is dated to c. A.D. 180 (Kenyon, 1948, p. 251, no. 1) while the two datable specimens of variety (66) were recovered from the Walbrook streambed, and is therefore pre-A.D. 155, and an early second century context at Newstead (Curle, 1911, pp. 324-325, pl. LXXXVII, no. 26). There is one dated example of the 'winged' S(ii) type (67), from an interesting context. It is assigned to the second century (Hunter and Kirk, 1952, p. 53) though the soil whence it came had been deposited post-A.D. 315. (*ibid*. pp. 46-48).

The distribution of the group is similar to that for the R brooches. Both for the group as a whole and for each subgroup individually (e.g. variety (64) see Richardson, 1960, fig. 3) there is a preponderence in the south rather than the north. There is no direct evidence for any manufacturing site.

Dolphin brooches

The dolphin or Collingwood's group H (Collingwood and Richmond, 1969, p. 295, nos. 12-17) is a common British brooch, the majority of which are unenamelled. The bow is humped forward over the junction of the arms. Enamel is however found on seven examples, in all cases on the bow. Blue is most common with one instance each of red and orange respectively and some decayed enamel still present. There are various cell shapes but only two are noteworthy. On the brooch found on Lambay Island, off the Dublin coast, (Jope, 1951, p. 60) the upper bow has a large cell (35 mms. x 4 mms.) with a bronze wavy line reserved along the centre and in which there is opaque medium blue enamel. A somewhat similar brooch from Charterhouse has two wavy lines in the cell but has now lost its enamel. Most have a simple one colour to a cell enamel but one has a small rectangular cell in the upper bow in which there are five small pieces of enamel now decayed/gone but obviously once two colours. Once more none comes from the north and the majority are from contexts datable to the late first to early second centuries though that for the Old Winteringham specimen is third century (Stead, 1976, p. 198, no. 18).

Knee brooches

Collingwood's group V or 'knee brooch' is a small brooch in which the head of the bow is like a bent knee with the lower leg for a tail. It is said to have been made in Britain in large quantities during the third century (Collingwood and Richmond, 1969, p. 298, nos. 82-8). Nine specimens with enamel have been encountered. Four of these have enamel on the semi-circular plate which acts as a spring-cover while the rest possess it on the bow. Blue and yellow are used as also turquoise and white (one intended, one decayed?) in addition to green (but also probably a decayed colour). The main cell shape is triangular though on three of the plates there is but one semi-circular cell. In all cases the enamel is one colour to a cell though in the triangles two colours alternate. Two brooches come from the north; the rest from the south of the country. One was found at Campsfield deposited in the fourth century but said to be second (Hunter and Kirk, 1952, p. 58) while another, in the Victoria and Albert Museum, was discovered near Eden Hall, Cumberland, along with coins of Hadrian and Antoninus Pius.

Thistle brooches

Collingwood's group W or thistle brooch is P-shaped in profile with a broad ribbed bow separated from a flat tail by a round or lozenge-shaped plate. There is a variety in which the bow is suppressed and the brooch has a flat plate with applied rosette, often enamelled. The type is commonly found on Claudian sites in Britain (Collingwood and Richmond, 1969, p. 293, nos. 89-90). The only enamelled specimens noted are those from Colchester where ten are stated to have been found (Hawkes and Hull, 1947, pp. 315-316). Most of these are said to have simple red enamel in the grooves on the bow and foot while a few are supposed to have a spot (or stud) of red enamel in the rosette. An examination of the objects in the Colchester and Essex Museum shows no positive traces of red enamel in these grooves. There are instances of what appears to be dirt which could have been mistaken for decayed enamel but it is now suggested that it never was enamel. There is however one example (reg. no. 31/31) of an applied rosette with a hollow cup as the head which is filled with red enamel. All these were found in contexts which are Claudian or just a few years later.

Group X as no. (96)

Group X basically consists of a bow and fan-tail (Collingwood and Richmond, 1969, pp. 294-295, nos. 92-7) and for the most part contains no enamel. However one variety as (96) which is a P-shaped brooch with a fan-tail has produced nine examples made to hold enamel. On all these the enamel is found in the fan-tail but there is one occasion when it also occurs on the bow in a row of small squares. The fan-tail is usually taken up by one cell in which there is reserved a bronze pelta-shaped design. For the most part the enamel is now decayed or entirely missing but red and blue have been noted - both simple enamels. Only one of these enamelled brooches comes from the north and none is closely dated though Collingwood (*ibid.*) places the group in the second century and one has been found recently in the filling of a fourth-century grave at Thistleton (Butcher, 1977, p. 64, fig. 10, no. 29).

Miscellaneous

Mention should perhaps be made of a group of less than twenty brooches which do not readily fit into any of Colingwood's groups. Some at least may be imports and therefore not relevant. In any event none is important or unusual enough to merit being noted separately. The enamel is all on the bow on which there is a great variety of cells, all geometric, as varied in shape as the bows themselves. Blue is the most common colour surviving followed by red and yellow; possibly green once existed but this may be another decayed colour as a lot of the enamel is now in this state. All of it is simple with the exception of three cases of a complex type - there is no millefiori decoration. Dates and distribution are of no consequence here.

This now completes the survey of bow brooches bearing enamel decoration. The remaining groups listed by Collingwood appear to be plain.

CHAPTER 3

ENAMELLED PLATE BROOCHES

There is an almost equal number of plate brooches from Britain but with these comes a greater problem as to whether they are of Romano-British manufacture or not. There is no doubt with the well known dragonesque brooches and the same may apply to what are here termed domed plate brooches as well as the 'oval and medallion' variety. As regards disc brooches (simple flat ones) and animal plate brooches the position is not so clear though many of these probably were manufactured in Britain. In this respect one might quote Wheeler, "It may be that the small animal-brooches found in duplicate near Mainz (Kleinwinterheim) and in London (q.v.) were made in Britain, though a claim might be urged on behalf of the enamel-workers of Belgica (p. 98). Here, as often, in spite of the resurging individuality of the British craftsman, it is difficult to draw a frontier line in Roman times between the cultures of southern Britain and north-eastern Gaul" (Wheeler, 1946, p. 34). In the last resort it may be difficult to be sure where these were made but this applies to the Continent as much as to Britain. Many of the miscellaneous disc brooches, the lozenge-shaped plate brooches and the various minor groups ('miscellaneous') are more likely to be Continental products and are therefore treated more scantily.

Drangonesque brooches (Figure 4A)

The dragonesque brooch is perhaps the most studied of British brooches. First listed almost forty years ago (Bulmer, 1938, pp. 146-153) they were later revised (Feachem, 1951, pp. 32-44) and even more recently have been brought up to date (Feachem, 1968, pp. 100-102). Collingwood has more to say on this type than on any other plate brooch (Collingwood and Richmond, 1969, p. 295, nos. 111-114). While Feachem's first article is the most useful he has little to say on the actual enamel though he sub-divides the specimens on the basis of the arrangement of their enamelled cells. This is remedied to some extent in his addendum where at least the enamel is described. Collingwood mrerely notes that some examples are enamelled. Since Feachem deals with all dragonesque brooches it is possible for once to know the proportion of enamelled specimens to plain ones.

Basically the brooch is a S-shaped plate with terminals so zoomorphic that it takes on the appearance of a sea-horse or dragon. The decoration is always on the front or top. Seventy-four of these dragonesque brooches are listed by Feachem and of these sixty-two or about eighty per cent are enamelled[54]. It might be asked, in view of the fact that the majority are intended to hold enamel, if some of the plain ones are not in fact un-finished (unenamelled), especially one group illustrated (Feachem, 1951, p. 43, fig. 6) of which two come from Traprain Law alone. Among the best know examples of the dragonesque type are the pair from Faversham in Kent (Feachem, 1951,

p. 39, nos. 4 and 5, fig. 3).

The enamel always occurs on the fatter middle portion of the brooch and the enamelled panel is therefore a rather fat S-shape with almost no curles. This is usually divided into three zones, the middle panel and two curving V-shapes at either end. As noted above Feachem divides the enamelled brooches into four groups on the basis of the shapes of the cell, viz. (i) the circular motif; (ii) the lozenge motif; (iii) the panel of squares; and (iv) the row of squares (figure 4A). With subgroup (i) the centre panel is circular, made up of a spot or spot and ring of enamel or with the circle broken up into smaller cells in various other ways; the end cells are divided into two or more cells, also curved (figure 4Ai). In variety (ii) the central panel is rectangular made up of dominant lozenge-shaped cells - varying from two to five in number - with attendant triangles; in some cases where there are three rows of tiny lozenges the panel may have as many as twenty-three minute enamelled cells; here the side panels are generally more elaborately split up with the septa taking on a much more 'celtic-type' appearance (figure 4Aii). The central panel of subgroup (iii) consists of a group of simple squares, four to eight in number, and again the V-shaped panels at either end are divided into two or three simple cells (figure 4Aiii). The whole panel (centre plus sides) in type (iv) is just a row of simple squares, usually five, with the end two slightly curved; sometimes the squares are fewer in number and rather crude (figure 4Aiv). The first two subgroups are more numerous than the last two.

Enamel is also generally found on the terminals of all subgroups. There is often a spot in the snouts, a spot or spot and ring as the eye and sometimes a cell on one or both sides of the central ridge of the ear. The terminal decoration is best seen on subgroup (ii) and is almost non-existent on subgroup (iv).

Red and blue are the colours most frequently met with, being clearly visible in over twenty instances each. Yellow occurs commonly, being present on over a dozen specimens. There is the occasional case of green and white though again it is not certain that these are always intentional and may merely represent the decayed remnants of one of the more usual colours. For example on one of the dragonesque brooches from Traprain Law the enamel is described as "?green" but seems rather to have the appearance of faded blue (Burley, 1955, p. 159, no. 33). There are over a score of specimens where part or all of the enamel has decayed but only a few where the cells are now entirely empty. There is one interesting specimen of subgroup (iv) from Caernarvon on which the few pieces of decoration are said to be alternately red enamel and pieces of chalcedony (Wheeler, 1923, pp. 133-134, fig. 56,3)[55]. Red, blue and yellow are commonly found together especially on subgroup (ii) but also appear on subgroups (i) and (iii). Red and blue however seem the more usual colour combination on these latter two varieties. Some examples of red and blue have been noted too on subgroup (i) and appear to be usual for the fourth group (iv) though on the latter red and green may also be present. Blue on its own was noted twice.

On all the dragonesque brooches simple enamels only were applied. While two or three colours usually appear together each is always confined to its own cell. With this group there is little doubt that the majority of finds are concentrated in the northern part of Britain between the Forth and Wroxeter. The Settle Caves, York, Corbridge and Traprain Law between them account for a third of the known finds. The group is assigned to the latter half of the first century and most of the second. Feachem notes the Wroxeter specimen found with coins of Hadrian (Bushe-Fox, 1916, p. 24, no. 9) and the Flavian example from Carlisle (Feachem, 1968, pp. 101-102). There is also an early second century find from Caernarvon (Wheeler, 1923, pp. 133-134) and the dragonesque brooch from the baths at Newstead (Curle, 1911, pp. 320, 332). There is therefore no real evidence for a date after A.D. 150 though the dated specimens are so few that this should not be ruled out.

Domed brooches

Among the disc brooches noted by Collingwood (Collingwood and Richmond, 1969, p. 229) one stands out and deserves special attention since it is not only invariably enamelled but seems to be without doubt a British product. The brooch (ibid. no. 102) consists of a circular hollow dome with a projection at the head (usually as a head loop) and one at the catch-plate and often two or more smaller knobs on the edge between (Figure 4B). This may be termed the true domed or smaller domed plate brooch to distinguish it from a larger version, of the same general type but different in terms of size and cell shapes (slightly more varied in design) as well as being numerically smaller. Neither seems to have been dealt with before as a type in its own right.

The smaller domed plate brooch (Figure 4B) is well exemplified by the three specimens from Woodeaton (Taylor, 1917, p. 115, nos. 65, 66 & 67, fig. 13)[56]. In all over forty of these brooches have been noted. The enamel is found on the convex surface of the dome on all examples and in a few rare cases also on the knob or projection at the 'foot'. The enamelled cells consist of two rows or circles of triangles and a spot or spot and ring on the top of the dome. The triangles point upwards and the bronze left between them (the septa) are triangles pointing downwards hence sometimes giving rise to the term 'zig-zag enamel decoration'. There are usually fourteen triangles in each row though thirteen and sixteen have been noted and a few cases of three rows of nineteen, sixteen and eight were met with. When it is remembered that the average diameter of the base of the dome (or the edge) is approximately 2 cms. it will be seen that the size of each cell side is therefore only 1-3 mms.

Blue is by far the most common colour found on the small domed plate brooches occurring on over thirty examples; this seems to be intentional as much as being due to a better survival rate. Red occurs on about half this number but more may have been lost due to weathering or decay. On approximately a dozen specimens blue and red appear together but there are cases of all the cells containing blue. Green appears on a few of these brooches for example on that from Camelon where it is found in combination with red (Anderson, 1900, p. 405). Black has been noted once and yellow is said to have been used on the example found at Yewden Manor (Cooks, 1921, p. 192); red and blue are also mentioned as being present.

On approximately a dozen brooches the enamel is too decayed to be able to say what the original colour was but on only three has all the enamel disappeared.

Each brooch has only a single colour enamel to each cell but generally two colours appear alternately though sometimes only one is used and there is the slight possibility of three. For this group the technique is always simple.

There are half a dozen dated specimens. That recovered during the excavations at Alchester comes from a level placed between A.D. 80 and 130 (Hawkes, 1927, p. 181, no. 3). The example from the Burgate Street excavations in Canterbury belongs to mid-second century levels (Williams, 1948, p. 20). What appears to be this type of brooch is said to have been found associated with a coin of Septimius Severus at Ixworth in Suffolk (Fox, 1900, p. 151) while two coins of Vespasian were found with the specimen from Wroxeter (Bushe-Fox, 1913, pp. 26-27, no. 9). The coins found at the West Park, Rockbourne, villa, which also yielded a small domed plate brooch, range in date from the mid-third century to the time of Arcadius (Hewitt, 1968, pp. 31, 36). This type of brooch would thus seem to have made its debut in the late first century and to have survived throughout the second and quite possibly into the third. Collingwood suggests the second to third centuries for its manufacture.

Few can be said to come from the north of the country. Thirty come from what can be termed southern Britain and half of these can certainly be said to have been found in the south-east.

<u>The large domed plate brooch</u> differs in various ways from the previous type: it is larger with a diameter varying between 3 and 4 cms; the dome while having a smaller base is usually higher; there is a brim or lip running around the base of the dome; there are generally more knobs on the edge and sometimes this is scalloped; overall the cell shape continues to consist of rows of triangles, found invariably on the dome. The type does not figure in Collingwood. Approximately twenty brooches belong to this group which is well represented by that found at Newstead (Curle, 1911, p. 331, no. 6, pl. LXXXIX, 20) and the specimen from Thistleton (Butcher, 1977, p. 53, fig. 6, no. 8).

Again it is enamelled on the dome but also sometimes on the brim and occasionally on the 'foot' knob. There is a greater variety of cell shapes though triangles are most common, but larger and fewer in number. On this group the instances of red and blue are about equal and this colour combination is the most common. There are a few cases of green if this indeed is the intended colour, and black is suggested as the second colour on the specimen from Ffrith (Davies, 1949, p. 247, no. 7) but this should perhaps be regarded as 'decayed'. Among all the examples looked at there are half a dozen instances where the enamel is now decayed and an equal number from which it has gone entirely.

As with the previous group all the large domed brooches have simple enamels with one colour per cell. Generally however two colours are used per object and the most frequent two are red and blue as already noted; red and green also appear and there is the odd case of only one colour, blue, having been applied. Two come from second century contexts at Ravenglass and Watercock (Olivier, 1979, p. 67, no. 2, fig. 26,2 and p. 211, no. 16, fig. 84,16). As regards distribution few have northern find-spots while further south the south-east has yielded slightly more than the south-west.

Oval and medallion brooches (Figure 4C)

This group consists of a flat oval plate with a separate small oval placed in the centre and rising a few millemetres above the surface of the former (figures 4C). The type is not listed by Collingwood but was briefly discussed when the specimen from Campsfield was published (Hunter and Kirk, 1952, p. 60). This example from Campsfield is typical of the type of which twenty-four have been noted.

The raised oval (equals the 'medallion') when it is present is always enamelled while the plate proper has one uninterrupted band of enamel running around the base of the medallion. The plate is usually 2.5-3 cms. long and about 2 cms. wide while the medallion is just over 1 cm. long and just less than 1 cm. wide; the width of the enamelled band on the plate is 3-4 mms. In only a few cases does the medallion still survive along with the main base plate and there appears to be no example where the medallion alone survives.

The enamels in the band on the plate and on the medallion are rather different. In the former blue and white are by far the most common colours and usually occur together though two cases of blue alone were noted (e.g. Brassington, 1980, p. 18, fig. 8,d). Red does not appear very often and then only with blue except in one instance green[58]. There is a small number of brooches where the two colours are decayed or even entirely gone. In the band the enamel is found as separate alternately coloured pieces usually squarish in shape. Twelve is the usual number but eight are also known and the edges between the pieces are quite clear, not blurred, though no metal separates them. This provides a good example of a more complex enamel-working technique. The enamel in the medallion is a simple one colour affair either red or green (the original colour in some cases at least) and usually bears an intaglio-like impression (incuse) of a human head as on those from Silchester (Reading Museum reg. no. 03291) or Wickham Bushes (University Museum of Archaeology and Ethnology, Cambridge) though there is the palmette based design on the Campsfield specimen. (Hunter and Kirk, 1952, pp. 59-60, fig. 26,1).

Most of these brooches are stray or unstratified finds but the few dated ones are interesting. That from Richborough is said to be "probably third century" in date (Bushe-Fox, 1949, p. 117, no. 48). The specimen from Silchester is assigned to the fourth century (Boon, 1974, p. 134, fig. 19,3) and the excavator suggests a fourth century date also for that from Lowbury Hill (Atkinson, 1916, p. 35, no. 34). We have seen (supra) that the finds from Campsfield come from an early fourth century deposit. There is thus no clearly dated early examples and contexts would suggest a third to fourth century date for these brooches.

There is only one specimen from the north (Corbridge) and none that can really be said to be from the south-west. Their distribution lies in an extended south-eastern area. While there would seem to be some Continental parallels[59], we appear to be dealing with a British brooch which shows a complex technique being made in the latter half of the Roman occupation and with a restricted distribution.

Chatelaine brooches

Another group of plate brooches which appears to be of British origin takes the form of a shield with a bar across the end held by two lugs and holding up to seven small toilet implements. The type rates only a line in Collingwood (Collingwood and Richmond, 1969, p. 300, fig. 106, no. 107). A fine example was discovered in the Romano-British cemetery at Baldock and is now in Letchworth Museum (Westell, 1931, pp. 261-262). Eight of these brooches have been noted to which may be added specimens from Wanborough and Gloucester (Butcher, 1977, pp. 51-52, fig. 5,5). The plates are invariably covered with a large number of small cells: rectangles, triangles (straight and curved), circles, petals (ovals) and shield-shaped cells, all containing enamel. This is always one colour to a cell and is generally blue, yellow, red or green and has a good survival rate. Most of the finds have been made in the south except for one from the Midlands and another from the north. That from Baldock comes from a late first century burial while another was retrieved from the Walbrook streambed (Guildhall Museum, 1960, p. 4, no. 4).

Similarly decorated, and therefore suggesting a common origin perhaps, are the small number of pelta- and crescent-shaped brooches such as those from Woodeaton (Kirk, 1949, p. 14, no. 48, fig. 3,5) and London (Wheeler, 1946, p. 96, nos. 35 and 34, fig. 29). Type 18 of the disc brooches, dealt with below but see a second specimen from Richborough (Cunliffe, 1968, p. 88, no. 6, pl. XXXI, 66), basically belongs here too on account of its decoration and indeed the centre of the plate is slightly domed not flat as with the majority of the disc brooches as defined. The decoration on this group and on the chatelaine brooches has most recently been described as 'scale-patterned' (Butcher, 1977, pp. 51-52).

Disc brooches (Figure 5)

Collingwood includes a great variety of plate brooches under his 'disc' type (Collingwood and Richmond, 1969, nos. 101-5). The domed type (no. 102) has been dealt with and that with the stud in the centre (no. 103) is discussed shortly. The variety with a conical stone set in the centre (no. 104)[60] and that with an applied repousse disc (no. 105) do not bear enamel. Only his (no. 101) is treated as a true disc brooch here, that is to define this type as a flat circular plate with nothing else except the brooch fittings (hinge/spring and catch-plate) on the under-side. The enamel is sunk into the upper surface in a great variety of cells. The various types in figure 5 are based on the design of the septa separating and defining the enamelled cells and are dealt with in this order.

In variety 1 the cells are circular with (a) one large cell covering the whole surface; (b) two cells; or (c) a spot in the centre and two circular cells around it. There are three examples of 1a completely filled with millefiori enamel. The specimen from Richborough provides a good example (Bushe-Fox, 1949, p. 117, no. 47, pl. XXIX, 47). It is not always clear if the two cells of 1b are intended for enamel and if so in a simple fashion as on the brooch from Boxmoor where the enamel is now rather decayed (Ashmolean Museum, reg. no. 1927.257). One from Wroxeter has a more complex technique of two enamels (one with additional spots in it) with no minor septa (Bushe-Fox, 1913, p. 23, no. 2). The third variety 1c is one of the largest with over a score of examples. The largish central cell is nearly always empty and may have had, in some cases, a stone rather than enamel as on a specimen from Carrawborough in Chesters Museum (reg. no. 2030). The majority possess simple enamel in the middle cell, usually red but an instance of green was noted, as well as in the outer cell which usually contains blue enamel. In one of the brooches from Corbridge (site museum) the blue is translucent not opaque as normal. There are some examples which do not retain their decoration at all. There are two cases of a complex enamel having been applied as well as three rather poor instances of millefiori decoration. Only the specimen from Caerwent is well dated being associated with late third to fourth century coins (Nash-Williams, 1930, p. 279). About twice as many have been found in the south as in the north.

Variety 2 is similar to the last (1c) except that in the middle cell there is reserved a circle of seventeen or eighteen bronze spots. On the half dozen examples known most of the enamel (all simple) is decayed or gone. Red, blue and green were however noted. The latter is to be seen on an example from Newstead (Curle, 1911, p. 331, no. 4, pl. LXXXLX, no. 6).

Type 3 differs from the previous group only in having a 'wavy' edge to the outer cell. On the nine noted the norm seems to be simple enamel with blue in the outer cell (and surviving more often) and red in the middle cell. See the two examples from Newstead (Curle, 1911, p. 331, nos. 3 & 4) and the one from a late fourth century context at Ravenglass (Olivier, 1979, p. 67, no. 5, fig. 26,5).

Variety 4 has a centre spot and three more cells, the outermost with reserved bronze spots in it. On the few noted most of the enamel has disappeared but green is present in the outer cell of that from Brough (Smith, 1866, pl. I,4).

About a dozen specimens of variety 5 are known. The specimen from Woodeaton (Kirk, 1949, p. 12, no. 25) is typical; a large central cell and one wide circle within which are six (in other cases eight) rings of bronze with a bronze spot in the centre of each. The centre spot has red enamel where this survives. The ground of the outer cell is usually blue (though rather weak and often tending towards green or white) but in a few cases actually green. The small circles of enamel appear to be red. A lot of the enamel is decayed but whatever the colour, it is always applied simply. No example comes from a dated context. A third are from northern locations.

There is only one example of type 6, from Pakenham in Suffolk (Ashmolean Museum, reg. no. 1927, 263). Like group 1 it has only one large cell but here are reserved one central and six other bronze spots. The enamel has however gone entirely.

The main septum in type 7 is star-shaped thus giving such a cell in the centre and six triangular ones on the edge which is 'wavy' as on the Brough find in the Ashmolean Museum (reg. no. 1927.302). None of the small number of specimens is alike in colour combination but red, blue and green are present in a simple manner. There appears to be no stratified find.

There are half a dozen brooches known with a zig-zag septum constituting type 8 (e.g. from Leicester in the Jewry Wall Museum, reg. no. 5541955). There is a ring of metal in the centre and the middle cell has a row of reserved metal spots. Most of the enamel is decayed or missing but simple blue (outer cell) and red (inner cell - with centre spot always empty)as on a specimen in the Ashmolean Museum (reg. no. 1927.249) seem to be the normal combination. However there are two cases of blue in both cells and in one instance it is translucent not opaque. Only that from Richborough is dated, seemingly to the late third or early fourth century (Bushe-Fox, 1949, p. 116, no. 45). The majority were found in the south.

Group 9 is denoted by a single triskele-shaped septum as with an example from Lowbury Hill (Atkinson, 1916, p. 35, no. 29, pl. IX, no. 29). Again only about half a dozen are known usually with red in the outer cell and blue within the triskele but blue also occurs in the former and both green and orange in the letter. In all cases a simple enamel with two colours is used. Most come from the south, all from undated contexts.

One of the most numerous groups is number 10 (including 11) with about two dozen examples. Again there is but one septum rather 'scalloped' with the scallops joining in six prominent knobs. Brettenham again provides a classic example of the group (Ashmolean Museum, reg. no. 1927.252). There is always only one colour to a cell, two colours to a brooch, in seemingly simple fashion. Red in the outer and blue in the inner cell is common but this can be reversed and there are instances of green, yellow and possibly white. In many cases the enamel is decayed or missing. One small group, 11, stands out as having black in the outer cell and either blue or red within the scallops as on that found at Wilderspool (Thompson, 1965, p. 83, Fig. 20, no. 9). Only a small minority have find-spots in the north and like those from further south are undated.

Group 12 has only a single example. Here the six knobs of the septum appear as the 'crests of waves'. The outer cell is blue and the inner red - the very centre is empty now. The brooch was recovered at Lowbury Hill (Atkinson, 1916, p. 35, no. 30).

On the six specimens of group 13 there are four semi-circular septa on the edge thus giving a cross-shaped central cell and four semi-circular ones between the 'arms' as seen on the Lombard Street brooch from London (Wheeler, 1946, p. 96, no. 36). The enamel seems to be of two colours, simply applied, per brooch: there are thus twelve instances of enamel but nine of these can only be described as decayed. There are two cases of red in the cross and one of blue in the arms. Five come from London or north-east of it, one possibly from Charterhouse.

Groups 14 and 15 are variations of the last type. The former has three semi-circles but no specimen still retains its enamel while the latter has six semi-circles seemingly blue, with the main cell filled with red. Altogether there are few specimens of these two types.

A few miscellaneous brooches make up groups 16, 17 and 18 of which the latter with thirteen specimens recorded is not strictly as defined here since it has a slightly domed centre as seen on the unstratified example from Richborough, on which the enamel has decayed (Bushe-Fox, 1932, p. 78, no. 14, pl. X, 14). The enamel, always simply applied, has frequently disappeared.

Overall the disc brooch constitutes a large group containing over one hundred specimens concentrated in the basically similar groups 1-5 and groups 9 and 10. Red and blue enamel are the most frequent but green can be said to be common and there are small numbers of black, white, yellow and orange. Simple enamel (usually with two colours) is the most frequent technique employed, but there are some cases of a more complex enamel and even millefiori decoration is present. Only about a fifth of the group comes from the north and very few are dated. The question as to whether these brooches are of British manufacture or not has already been touched upon in the introduction to this section.

Animal plate brooches (Figure 6)

This group is as numerous and as varied as the last. The majority are flat though some are in the round such as the cock/hen brooches. Collingwood deals in three lines, and without real discussion, with what he terms the zoomorphic plate brooch (Collingwood and Richmond, 1969, p. 300, nos. 106, 109 and 110). Again each type is discussed separately in conjunction with the illustrations provided in Figure 6; the order follows the type of animal represented in alphabetical order. The enamel is always on the upper surface. Some representational forms other than animals are included here such as the axe, horse and rider and sandal.

There are very few examples of form 1, an axe, but that from Camerton (Somerset County Museum, Taunton, reg. no. A 101) has a simple enamel in blue and a now decayed colour.

Under group 2 comes a number of miscellaneous bird shapes which do not readily fit into the ducks, eagles and cock/hens dealt with below. They are variously described as ducks, pigeons, doves, peacocks and pheasants and most are partly in the round (i.e. head only) as the 'pigeon at rest' from Chanevage, Gloucestershire (Ashmolean Museum, reg. no. 1927.381). The simple enamel is generally found on the body and wings and where it remains clear is blue and red.

There are a few brooches in the shape of a boar, type 3, one with blue enamel on the body having been found in a late second-early third century deposit at Brough (Corder, 1935, pp. 32-33).

Group 4 is one of the largest and best known types being a sitting (?roosting) cock or hen in the round. Over a dozen are known among them the well published specimens from Bowland (National Museum of Antiquities of Scotland reg. no. FA87-1922.521) and York (Yorkshire Museum, reg. no. H 139A) both enamelled in red and yellow. The enamel is found in small simple cells on the body and sometimes the head. Much of it has decayed or is missing but some is red, yellow and blue, usually two of these together - red and yellow, blue and yellow as well as blue and red. The enamel is simple but usually well executed. The majority of these brooches comes from the south and only one is dated, that from Wroxeter found in a context dated before A.D. 150 (Bushe-Fox, 1916, p. 25, no. 13).

A small number in the shape of a deer, group 5, was noted. That from Caerleon (site museum) has one cell covering the body - this has a complex enamel of a blue ground with three yellow spots inserted while the specimen from the Well of Conventina at Carrawburgh (Chester Museum, reg. no. 2033A) has a simple enamel in red only.

Dogs constitute group 6 and of the dozen noted the most common is what is possibly a hound running to the right - sometimes erroneously called a hare. The main cell covers the body but occasionally the eye is enamelled also. Some are simply decorated in blue where the enamel survives, others have a complex decoration of a bed of blue into which spots of yellow are set - red with white spots is also known. None is dated and the majority were found in the south.

Group 7 is small. It consists of dolphin-shaped brooches such as the example from Newstead (Curle, 1911, p. 332, no. 1, pl. LXXXIX, no. 3) which has a simple enamel decoration in blue and a second colour now decayed.

Lincoln provides a specimen of the group 8 'swimming duck' type (Thompson, 1954, p. 127, fig. 4,8) with a flat body and a head in the round. Simple enamel is found on the back and wings; rarely do the eyes have settings. Of the two colours used one is usually blue, the other, where it survives, red. Most have been found in the south of the country. The two from York come from a grave group containing a coin of Trajan (Royal Commission, 1962, p. 91) while another comes from the Walbrook and the Saint Albans specimen was found in a mid-second century context (Frere, 1972, p. 117, no. 21).

A small number, group 9, take the form of a perched eagle (sometimes described as a crow). Mostly the enamel has decayed or disappeared but that from Castlelaw retains its red and black enamels; the date suggested by the excavator is the second century (Childe, 1933, pp. 383, 386).

Fish too appear in the zoomorphic repertoire as type 10. Just over half a dozen, all flat, are known such as the two found in the

Walbrook streambed (Wheeler, 1946, p. 98, no. 40 and Guildhall Museum, 1960, p. 4, no. 1). The enamel, always simple, is mostly decayed or gone now and no example comes from the northern parts of the country.

Groups 11 (fly), 12 (frog) and 13 (hippocamp) have few representatives. Only the first has simple enamel, the remaining two possess a more complex technique. The frog brooch from Leicester (Jewry Wall Museum, reg. no. 116, 1962/872) has two cells on its back, both with a decayed ground containing four white spots while the hippocamp from Silchester (Reading Museum reg. no. 03282) has a red ground.

Another numerous type, 14, takes the shape of a horse and rider (flat and to the right). On over a dozen examples the enamel is found on the bodies of the horse and the rider - as on the pair from Woodeaton (Kirk, 1949, p. 14, no. 30). The enamel seldom survives well but blue and red appear to be the normal pair of colours with the former still preserved more frequently. It is always applied simply. No less than six were recovered at Hockwold, Norfolk, from a possible fourth century context and four more have been found on the site of the late temple on Lamyatt Beacon, Somerset (Butcher, 1977, p. 54).

Almost as many horses alone, group 15, have been found though here the variety of shape is greater. Only one comes from the north namely the specimen from York (Yorkshire Museum, reg. no. H1398). The majority possess simple enamel, the colour mostly not distinguishable now but blue and green were noted. None is closely dated.

A few of these animal brooches take the form of a leopard (type 16) and a lion (type 17). The enamelled spots - mostly empty now - on the example from Harwich (British Museum, reg. no. 81.9-9.19) probably denote the former.

Another of the larger groups, 18, denotes a hare or rabbit - sometimes obvious, sometimes not clear, which. Such a case is the early third century brooch from the Jewry Wall site in Leicester (Kenyon, 1948, p. 251, no. 5). Again the enamel is mostly simple but usually in poor condition though blue and red have been noted. The majority come from the south where Winchester yielded a specimen lost about A.D. 60 (Hull, 1964, p. 90, no. 5, fig. 24,5).

Equally large is group 19 in the shape of a sandal as for example that from Wilderspool (Thompson, 1965, p. 83, fig. 20,7). Here however a complex enamel is always in evidence. The one large cell is filled with a blue (except orange in two cases) ground into which are placed spots of black, white, yellow or red. Only a small proportion have northern find-spots. That from Shakenoak can be dated to the second century (Brodribb et. al. 1973, p. 108, no. 179).

Finally group 20 is a stag such as that from a late second century context at Wroxeter (Atkinson, 1942, p. 208); this example has blue and green enamel.

If we take the animal plate brooches as a group the main technique is a simple enamel usually in two colours but complex enamels do occur not only on the sandal brooches but also on such as on the dog brooches; millefiori decoration is not met with in this group. There is a large number of decayed enamels and even entirely empty cells. As regards colour blue seems to be the most frequent, appearing over fifty times and being three times as frequent as red. Yellow, green, orange and black are found in smaller quantities; white appears as spots in some of the complex enamels. Out of over a hundred specimens less than twenty per cent can be said to have been found in the north. Too few are dated closely enough to attempt to place the whole group in any subdivision of the Roman period though generally second and third century contexts seem most common.

Miscellaneous disc brooches

There are almost seventy brooches of various shapes which come within Collingwood's disc brooch type but not that latterly defined. They are basically a circular plate but have additions such as projections and/or head loops on their edge and especially some sort of applied upstanding decorative rivet or stud in the centre. Here a British origin is less certain so the types are discussed only briefly.

There are about a dozen with no upstanding studs but with a head loop and from four to ten circular projections around the edge. Most have the enamel in simple circular cells though it is now in a decayed or worse condition but red, blue and green are present. There is one example of a complex enamel in orange and decayed square pieces and a poor but possible millefiori specimen. Only one was found in the north.

The majority have however some sort of separate applied rivet in the centre; mostly this is plain but sometimes it consists of a metal cone and in a few cases a zoomorphic shape in the round. There are nine examples with millefiori decoration such as one from Richborough (Bushe-Fox, 1949, pp. 116-117, no. 46). Of the remainder just slightly more have simple enamel: in red, blue, yellow, black and white though much cannot now be distinguished due to decay. There are many complex enamels in this group, the majority consisting of separate squarish pieces of two different coloured enamels laid side by side: blue with red, yellow or white; green with red or black, and red with yellow were all noted. Most cells are circular and narrow though of varying widths.

Very few of the find-spots are in the north and of the remainder the greater number occur in the south-east. Saint Albans has yielded one type from a late fourth century deposit (Kenyon, 1934, p. 259, no. 13), two more from an early second century context (Wheeler, 1936, p. 209, nos. 34 & 37), and a fourth associated with a coin of Domitian as well as Flavian pottery (ibid. no. 36). Two have been recovered from the Walbrook, Wroxeter has also yielded two early second century examples (Bushe-Fox, 1916, p. 25, nos. 11 & 12). A late third century specimen comes from Richborough (Bushe-Fox, 1949, pp. 116-117, no. 46). Thus while there appear to be late examples the trend is towards the second century.

Lozenge plate brooches

Here the shape is basically a lozenge, usually with projections at the angles but also with zoomorphic (often debased) projections at the 'head' and 'foot' on approximately a third of the forty or so specimens examined. The enamel is normally found within the lozenge in simple cells. Half these brooches have simple enamel mostly decayed or gone now but blue, red, white and green have been noted. Over a dozen have a complex enamel either pieces of blue, black, white or green or else a ground of white, blue or green with black or white spots. Two specimens with millefiori enamel were noted - one a rather plain lozenge brooch from Bardwell (Hills, 1871, pp. 258-259, pl. 12,1). The other zoomorphic lozenge brooch type is exemplified by one with simple enamel from Richborough (Cunliffe, 1968, p. 89, no. 67). The lozenge brooch comes overwhelmingly from the south. Only two possess dated contexts however i.e. the zoomorphic specimens from Colchester from a Trajanic level(Hull, 1958, p. 117) and Saint Albans from a mid-second century level (Frere, 1972, p.118, no. 23).

Miscellaneous plate brooches

In addition there are over a hundred other enamelled plate brooches from Britain, some of which are however well known 'foreign' types with zoomorphic terminals. Other brooch shapes include ovals, squares, rectangles, open circles (wheels), triangles and various other shapes, some flat, others stepped or raised, and several with projections, often zoomorphic. Many have simple enamels with blue, red and yellow very common and orange and green less so. A large number possess complex enamels made up of pieces of blue, red, white, orange, black and yellow or else a ground with spots in various combinations of these colours. Only three have millefiori decoration. Once more the majority come from the south and the few dated lie in the second century.

CHAPTER 4

ENAMELLED OBJECTS (NON-BROOCH)

Enamelled objects other than brooches are fairly numerous and make up about one third of all objects bearing this type of decoration found in Roman contexts in Britain. Seal boxes constitute by far the largest group providing almost two hundred enamelled specimens. Other types found in large numbers include button-and-loop fasteners, circular studs or mounts, miscellaneous studs and mounts, belt-plates and various other miscellaneous objects. Known in smaller quantities are vessels, miniature stools or stands and finger-rings. Only the button-and-loop fasteners have been previously examined in any depth (Wild, 1970, pp. 137-155) though much discussion has taken place on the vessels (see most recently Moore, 1978, pp. 319-326) and to a lesser extent the stools while various individual items such as the Thames plaque have received some attention (Brailsford, 1964, p. 56).

As regards what is of Romano-British manufacture and what is of Continental origin, what to include here, what to omit, much the same problems arise as with the plate brooches. Some are undoubtedly British, some foreign and many belong to types having a wide distribution and a home equally possible here or abroad. In this respect at least the Thames plaque in its discarded, unfinished state is certainly a home product and the button-and-loop fasteners found in their moulds at Traprain Law are further evidence of a British origin for this group. The miniature stools also seem to be a product of Roman Britain and some at least of the finger-rings, vessels and belt-plates fall into this category too. The seal boxes are rather akin to the animal plate brooches but again many may have been made by the Romano-British enamel-working industry. The studs and/or mounts are a problem but should not be dismissed as being foreign.

Button-and-loop fasteners (Figure 7A)

This is the only category comprehensively dealt with elsewhere. These objects were first listed as dress fasteners (Gillam, 1958, pp. 79-85, 90) but more recently have been assigned to the horse harness repertoire (Wild, 1970, pp. 137-155) [61]. The type consists of circular, rectangular or other shaped 'buttons', from the centre of the backs of which project a shank which then turns at 90° to lie parallel to the head; this shank is usually longer than the button and ends in a loop, often triangular. Eighty appear in the first list and one hundred and sixty-five in the second. The latter however includes a large number of Continental finds, some in bone, as well as those already discussed under the Iron Age enamelwork. There are almost sixty enamelled specimens, quite a few of which are not in Wild's list[62]. Approximately fifty per cent of the group bears enamel. Those listed (*op. cit.*) from Traprain Law provide a good selection of both enamelled and plain examples.

It is perhaps best to follow Wild's typology. Out of his ten classes only groups III, V and VI are of interest here (figure 7A)[63]. Although class III is numerically the largest group only a small number bear enamel decoration on their tear-drop or petal-shaped buttons. In three cases the central boss bears blue enamel but in the other three there are two colours, green and red in one instance but decayed or missing otherwise. Fasteners with circular buttons, sometimes convex, constitute group V: subgroups a, c and d are plain but Vb always has enamel, frequently in the shape of a flower. Wild lists only seven but over twenty were noted during research for the present work. Blue and red are the most common colours here with yellow less so and a few possible instances of white; now empty cells or cells with decayed enamel are frequent. Where only one colour is applied blue seems to be the choice; where two are to be seen red and blue are most common but blue with white and red with yellow also occur; there are a few cases of three colours viz. blue, red and yellow. Group VI is characterised by flat rectangular buttons - (a) being larger and squarish, (b) plain and (c) small and elongated. The cells are more geometric than in group V. For this work six of the larger and four of the smaller were listed. Red, blue, yellow and possibly green and white are the colours which are applied in pairs - some have now decayed or disappeared entirely. There are a small number of enamelled button-and-loop fasteners which are 'atypical' - the more noteworthy are dealt with in the chapter on Iron Age enamels.

In all cases the enamel is applied in a simple fashion, one colour to a cell, on the upper surface of the button; no complex nor millefiori enamel was noted. On the whole two colours occur together, most frequently red and blue; one and three colours also appear, small numbers of yellow, green and white occur; a fair proportion has now weathered. Cells tend to be small, being more geometric (square, triangle, lozenge) on the rectangular group but more curved (petals etc.) on the disc-shaped buttons.

By far the great majority were found in the northern parts of the country especially on military sites such as Corbridge, Newstead and York. However it must be noticed that the only evidence for manufacture comes from the native settlement on Traprain Law. As regards date most are unstratified or unassociated or possess no exact details of recovery such as with the six examples in Chesters Museum. Wild (1970, p. 146) does not believe that the type was manufactured later than the second century when the majority were probably made; he regards the two from Caerleon (*ibid.* no. 61) and Campsfield (*ibid.* no. 62) both with late contexts to be survivals.

Miniature stools or stands (Figure 8A)

A small group of objects about which relatively little has been written comprises the miniature stool or stand (Figure 8A). This consists of four legs, which widen out at the top to form four aches or faces, surmounted by a flat top. The top has a circular opening in the middle, a small knob in each corner and slightly in from each knob a hole. From the pairs found attached it is obvious that the stools are meant to join - the four legs of a smaller one fitting into the four holes at the corners on the top of a larger one. The legs of the complete lower one from Caerwent (Newport Museum) show it to be a basal specimen measuring 4.5 cms. high and the top 4 cms. square. The example in the Britiah Museum from an unknown location (reg. no. 57.12-18.1) without the four holes and with a funnel occupying the central opening appears to be a top example; it measures 3.6 cms. in height with top edges 1.6 cms. long. It seems that a complete set would thus consist of three stools diminishing in size from the lowest to highest. Their purpose is uncertain though the belief that they have 'ritual significance' is perhaps reinforced by the specimen from the Brigstock shrine (Greenfield, 1963, p. 243, fig. 4 ,3 and Taylor, 1963, p. 268) and the three from Farley Heath (Jessup, 1970, p. 195, fig. 58)[64].

Over twenty of these objects have come to light and they can be exemplified by the three pairs still attached viz. those from Caerwent (Newport Museum), Silchester (Reading Museum) and South Shields (Green, 1978, p. 72, pl. 128-9). Enamel is always found on the outer faces of the four sides or arches between the upper parts of the legs. A large amount of this has now weathered or in some instances has completely disappeared. However on the nineteen where one or more colour is discernible blue is present in each case. In some instances this is a rather greeny blue and in others it is decaying into a white substance. Green seems to be present as an intentional colour on a few occasions as also is yellow. Red occurs on about a third of the group. There may be the odd case of blue applied on its own but red and blue is likely to have been the usual colour combination. There is one instance of red, blue and green together. For the most part the cells on each face are three in number and the main designs are shown in figure 8A. In all cases the technique used is simple with one colour to a cell; complex or millefiori enamel is not present. The technique of notching some of the cell edges to facilitate the holding of the enamel on some of these objects may be noted.

Of the eighteen with known provenances seven come from the north , four of these from South Shields, two from Corbridge and one from Chesters. The remaining find-spots are reasonably well spread over the rest of the country and include three each from Caerwent, Silchester, London (Green, 1975, p. 65) and Farlay Heath (Green, 1976, p. 219). The only one clearly dated is that from the shrine at Brigstock dated by the coins post-260 to c. A.D. 380 (Greenfield, 1963, p. 243) though one from the 1908 excavations at Caerwent may have an association with a coin of Constantine Junior (Ashby et. al. 1910, p. 15). There may therefore be a late date for this interesting group of enamelled objects.

Seal Boxes (Figure 7C)

Although one of the most numerous types of object found on Romano-British sites seal boxes remain a neglected subject. The function is implied in the name and the method of use has been explained on a few occasions (e.g. Bushe-Fox, 1916, pp. 28-29, pl. XVIII, 25; Wheeler, 1946, p. 108, fig. 33)[65]. Enamel decoration is frequently found on the lids of seal boxes[66] and indeed in many cases only the lid survives though the same is also true of the box proper. It would require too deep a search to discover what proportion bear enamel but it was noted that only four out of the thirteen housed at the then Guildhall Museum had been so treated.

Almost two hundred enamelled seal box lids have been found in Britain. There are five basic shapes[67] of seal box and the enamel is discussed as it occurs on these (Figure 7C). Since it survives wholly or partly on about sixty per cent of the lids there is thus a reasonably clear picture of the colours and techniques employed. Approximately a quarter belong to group 1 which is basically tear-, petal- or pear-shaped. The majority have simple enamel in two colours, usually blue and red though yellow and green appear on a few, in various combinations of these four colours - a lot of the enamel is now decayed. Less than a dozen have more complex enamels equally divided between pieces of blue, red, yellow and white and beds of blue or green with red, yellow or white spots and sometimes with a further ring of enamel around the spots. Newstead (Curle, 1911, p. 330, no. 2) provides an example of the former as well as one of the complex variety (ibid. pp. 332-333, no. 5). Almost as large is group 2 with more or less circular shapes. Again most are simply enamelled in two colours out of red, blue, yellow, green and orange with the first two most common and usually together. Just over a dozen possess complex enamels usually a ground (blue, green, yellow) with inserted spots (white, orange, yellow) but less often pieces in alternate colours of blue, red, yellow, white and green. A simple example comes from Richborough (Bushe-Fox, 1949, p. 124, no. 78) and a complex one from Cramond (Rae, 1974, p. 194, no. 4, fig. 14,4) - alternate orange and blue pieces in a circular cell.

Over two dozen lozenge-shaped specimens constitute group 3. Half of these are similar in having the surface of the lid completely covered with small lozenge-shaped cells - twenty-five in number - giving a latticed effect. Where the enamel has not decayed it is normally blue or yellow and this may be the usually intended scheme though there are possible cases of yellow alone and one case of yellow and green. Chesters Museum possesses a good specimen of this class found at Great Chesters (reg. no. 2487; Budge, 1907, p. 396, no. 301). All of the latter variety have simple enamelwork as have all but three of those without the lattice design. Here the surviving enamel is red and blue and in the three complex examples the colours found are red, blue, white and green.

Group 4 is rectangular and out of the seventeen specimens known, eleven have a celtic-type design, in a simple enamel, of red and blue though there are rare occurrences of yellow and green and several cases where no enamel has survived at all. Examples of this type have been found at Chester (Chesters Museum, reg. no. 1911) and Lincoln (Brailsford, 1964, p. 78, no. 10, fig. 40). Where the cells are geometric they consist of small squares in rows (twelve or sixteen squares in all) with the enamel mostly gone now though orange and blue were noted. One with sixteen squares containing enamel, now largely decayed, comes from Burgate Street in Canterbury (Jenkins, 1951, p. 33, fig. 16,21). It is to the last variety that the two certain examples of millefiori enamelwork belong. No instance of a complex enamel is met with in group 4.

Finally there is a small group, 5, generally tear-shaped with an applied head (horse or cow though these descriptions are not always accepted) on the lid. There is usually a small setting in the snout which has a spot of red enamel in the few instances where it survives. Such a lid was retrieved from the barracks at Newstead (Curle, 1911, p. 333, no. 7).

Distribution as regards the individual groups and as a whole lies over the whole country with perhaps the south-east claiming a slightly larger number than the south-west or north. The majority have been found on town sites and military sites. Excavations at Corbridge produced fifteen enamelled specimens (now in the site museum) and Colchester has yielded no less than twenty. A greater number than usual come from dated contexts. One from Caernarvon was lost not later than c.230 (Wheeler, 1923, p. 141). The specimen from Cramond, noted above, was probably made during the second century. Canterbury provides an early example c.A.D. 75 (Jenkins, 1951, p. 33) as does Colchester c.A.D. 43-65 (Hawkes and Hull, 1947, p. 333, no. 7). A specimen from Fishbourne was dropped either in the second or third century (Cunliffe, 1971, p. 118, no. 129). Three and possibly a fourth were recovered from the Walbrook streambed. An example from Maxey was found associated with third to fourth century pottery[68]. One from Nottingham has a firm Flavian dated (Megaw, 1971, p. 153). Richborough yielded another from a context earlier than the end of the third century (Bushe-Fox, 1949, p. 124, no. 75). Another early specimen comes from a level dated to the first half of the first century A.D. at Saint Albans (Frere, 1972, p. 122, no. 67) while a second is dated to the mid-second century (Wheeler, 1936, p. 212, no. 52) and a third comes from the fill of a late third century pit (<u>ibid.</u> no. 53). Three from Wroxeter appear to belong to the first half of the second century (Bushe-Fox, 1916, p. 28; Atkinson, 1942, p. 211).

As a whole seal boxes appear to have been discarded/lost as early as the mid-first century A.D. and seemingly right through to the fourth although the second century seems to have seen them in greatest use. Bushe-Fox (1949, p. 124) has remarked that "they do not seem to have been recorded from late third to fourth-century sites". The techniques which are used are mainly simple though complex ones appear on about fifteen per cent and millefiori on two specimens at least. The enamel on the group 1 specimen from Leicester (Jewry Wall Museum, reg. no. 116. 1962/729) is noteworthy in that it has a ground of yellow speckled randomly with brown 'spots'. This is unusual but is met with again. One might also note the notching on the cell edges on the lid from Carmerton (Bristol City Museum).

<u>Vessels (Figure 8C)</u>

Some of the best known individual enamelled objects belong to this group. It comprises nineteen complete or fragmentary objects, for the most part either paterae or what are termed 'cups' though small bowls may be a better term. When only a fragment of the body remains it is difficult to place it in the correct class. There has been no detailed study of these objects but a certain amount of discussion has taken place (e.g. Franks, 1870, pp. 515-516; Curle, 1932, p. 306; Henry, 1933, pp. 110-128 <u>passim</u>; Cowen and Richmond, 1935, pp. 310-342; Butcher, 1977, pp. 45-49; Moore, 1978, pp. 319-326).

The paterae are dealt with first, then the cups or bowls and finally the miscellaneous vessels.

The paterae bear enamel on the outside of the bowl and on the upper surface of the handle; often the latter is all that survives of the vessel. The most outstanding complete example is that from an unknown location in West Lothian and now in the National Museum of Antiquities of Scotland (reg. no. FR43). The outside of the bowl is enamelled in a simple technique in five zones with red, blue and green. The main zone has a large running vine-leaf pattern, above and below which is a narrow band with a wavy line of bronze reserved in the middle; the upper and lower zones are respectively a row of double laurel leaves reserved and a row of vandykes; to be noted are the notched or toothed edges of much of the design. The handle, again in blue, red and green, has a basic design of two leaves. There is no definitive publication of this object though it has often been noted and is generally assigned, as a patera, to the second century (Anderson, 1884, pp. 44-45, pl. I; Curle, 1932, pp. 302-306, 353, fig. 14). A somewhat less pleasing example comes from a cremation burial in the cemetery at Brougham (Charlesworth, 1968, p. 306, pl. LXXVIII b). The handle is plain and is believed to be a later addition to what was primarily a cup or bowl (ibid.). Again it is simply enamelled in blue, green and red and has much toothing to the cell edges. A handle only survives from Kirkby Lathorpe, Lincolnshire, as a stray find (Butcher, 1977, pp. 45-46, fig. 1). The design is very similar to that on the West Lothian patera handle but here the enamel has mostly decayed though traces of blue remain. Handles only, with a somewhat different design, have been discovered at Brough-on-Fosse, Nottinghamshire, and Burgh Apton, Norfolk (Moore, 1978, p. 319, fig. 1 and p. 322, fig. 3). Both possess blue and green enamel with red and yellow as a third colour respectively; the outer edges of the Burgh Apton handle cell are toothed. Another handle, from Dinorben, has lost its enamel (Gardner and Savory, 1964, pp. 148-149, fig. 17). Mention may be made here of the patera found at Amiens with its cranallated design of numerous small geometric cells (blue, green, red) and the names of forts on Hadrian's Wall (Heurgon, 1951, pp. 22-24, fig. 4).

The Braughing cup (Franks, 1870, pp. 514-516) has a design similar to the West Lothian patera with a running vine-scroll, laurel leaves and vandykes; the enamel is only blue and green here and again toothing is much in evidence on the cell edges. The design on the Harwood cup is much more simple and rather unique and seems only to possess a simple blue enamel (Franks, op.cit. p. 515). One of the best known cups is that from Rudge with the Wall fort names and crenallated design of numerous small cells, rectangular and crescent-shaped, in red, blue and green (Cowen and Richmond, 1935, pp. 310-342)[69]. There is a fragment of a bowl with a somewhat similar design but without the names; this was acquired in Barcelona, is known as the Hildburgh fragment and is housed in the Victoria and Albert Museum (Cowen and Richmond, 1935, pp. 332-334, fig. 1 and pl. XXVIII,2) - a lot of the enamel is gone but blue and seemingly yellow and green are used. Another fragment with a basically similar design of small geometric cells comes from Great Chesterford and has only some traces of blue and decayed enamel remaining

(University Museum of Archaeology and Ethnology, Cambridge, reg. no. 48.964). The design on the fragment from Bradley Hill (Wilson, 1971, p. 276) is based on a series of five-sided polygons running round the greater part of the body; blue only remains for certain but there is possibly red and a third colour now decayed. A more or less similar design appears on a complete cup from Canterbury (Henry, 1933, p. 120, no. 2)[70]. Another fragment of a cup with inscription and scroll design in blue and green and with some notching of the cell edges has recently been recovered at the Beadham Villa in Yorkshire (Wright and Hassall, 1973, p. 334, fig. 21). Two small fragments but with no enamel remaining have quite different designs and were found at Halton Chesters (Henry, 1933, p. 142, no. 2) while Moore (1978, p. 322) mentions the further find of two enamelled bowl fragments at Bough-on-Fosse.

The famous Barlow vase though having a different function has similarities in design with the West Lothian and Braughing objects (Gage, 1836, pp. 303, 307, 311-314, pl. XXXV). The enamel was blue, green and red though on the original this is now a blistered mass having been in a fire. Despite the fact that the copies have a handle attached by escutcheons to the body the enamel continues unbroken around the original body. Both the latter and a copy are in the British Museum.

A two-handled vase in the Fitzwilliam Museum, Cambridge from an unknown location is different from the preceding objects (Dalton, 1912, p. 100, no. 44, pl. XII). Its enamel is red, green and orange.

Seventeen of the nineteen objects here may be classed as the patera/cup/bowl type simply enamelled all over the outside of the body (and upper handle where appropriate) in red, blue and green. The term 'skillet' is now generally applied to the paterae and it has been suggested that the cups or bowls are in fact skillets which have lost their poorly applied handles (Butcher, 1977, p. 49, and Moore, 1978, p. 319). If the Harwood cup is excepted there are three basic designs (see Figure 8C); that where a vine-scroll is prominent; small geometric cells in a crenallated pattern; and the polygonal type recently designated the 'Vehner Moor type' after Eggers (Moor, 1978, p. 321). The question of origin is difficult. Three workshops are possible but the use of notching/toothing on both the first and some of the third type (e.g. those from Namur and Rocheford - Henry, 1933, p. 120, nos. 1 & 3) is interesting. With the exception of a specimen from Jutland (ibid. p. 112, no. 4) all the first group have British provenances and were probably made here. The second group are obviously Wall souvenirs and though such objects could have been manufactured abroad (though too early for Hong Kong) a Romano-British origin is plausible. One may still suspect a Continental origin for the third group such as those from Canterbury and Bradlay Hill though the technique for helping to hold the enamel is similar to that on the first type. This agrees with Henry's views (1933, pp. 114, 122) though Franks (1870, pp. 515-516) prefers a wholly British origin and Curle (1932, p. 306) suggests a Continental one. The excavators believe the Brougham patera (Charlesworth, 1968, p. 306) and the Dinorban handle (Gardner and Savory, 1964, pp. 148-149) to be both of local manufacture. In her discussion of the group of enamelled objects Butcher (1977, p. 43) follows Henry in seeing a mixed British and Continental origin while Moore (1978, pp. 324-325) argues strongly for a wholly British manufacture. This is very attractive and indeed quite possible but one still wonders if the polygonal or Vehner Moor type can be ascribed to a Romano-British workshop.

Circular mounts (Figure 7B)

This section deals with mounts or studs consisting of a circular plate with a single rivet or stud at the centre of the back (Figure 7B). The plate varies in diameter being fairly evenly spread between 1.5 cms. and 5.5 cms; it is usually a few millemetres thick with generally a straight edge though some with slanting edges are slightly 'hollow'. The one stud in the centre of the underside is usually incomplete now but where it survives is t-shaped. About seventy are known and the type is well illustrated by the groups from Newstead (Curle, 1911, pp. 331-332, pl. LXXXIX) and Chesters (Budge, 1907, *passim*). They appear to be decorative mounts in some instances for horse harness though other uses are possible.

The upper surface of the plate generally contains enamel in plain circular cells. The great majority possess either two or three cells; a small number display four cells or at the other extreme only one; only half a dozen do not conform to the normal pattern. The width of the cells varies considerably.

Unlike most of the groups so far few of these mounts have the enamel applied in a simple fashion. There are a small number where red, blue, yellow or green is the only colour in the cell. One from Manchester has turquoise enamel within a triskele-shaped septum and a yellow colour in the outer cell (Jones, 1974, p. 123, no. 20). Approximately a third of the group has a more complex enamel applied in equal numbers as pieces or as a ground of one colour with spots of another inserted (many of these spots have a yet different coloured ring around them). Many of the pieces are now weathered but blue, red, yellow, green and orange appear in various combinations of two colours in squarish pieces laid side by side. In the type with spots blue is the most common colour for the ground though green and yellow were noted; the colours of the spots - and rings where used - are blue, white, yellow, red and black - white is most frequent and black often occurs as the ring. A good example of this is seen on the stud from Mumrills which in the outer cell has a light blue ground with spots and circles surrounding them (MacDonald and Curle, 1929, p. 555, fig. 115, 9). On a half dozen there is a combination of a complex and millefiori enamel - i.e. alternate plain pieces and pieces of millefiori. However most of the millefiori enamel is found on its own, on over a third of the specimens known. This is one of the largest groups of millefiori enamel. It might be noted that this form of decoration is more frequent on the larger mounts (4-5 cms.) which have no other technique applied. The specimen from Chapstow is a good example (Brailsford, 1964, p. 56, no. 6, pl. XXI,6).

Over fifty per cent of this group were found in the north, the rest are evenly spread over the southern parts of the country. It is interesting to notice perhaps that almost seventy per cent have been recovered from military sites; twelve from Chesters, eight from Corbridge, and six from Newstead. Despite such a large number only three possess dated backgrounds. That from Chesterholm was excavated from a fourth century deposit (Brailsford, 1964, p. 56, no. 5) while two come from mid-second century deposits at Saint Albans (Frere, 1972, p. 126). Unfortunately those from Chesters are as good as unstratified but the Newstead examples appear to be second century in date.

Miscellaneous studs or mounts (Figure 9B)

Included here are studs or mounts other than those simple circular ones with only one stud just discussed (Figure 9B). Studs - usually two, sometimes four - are still present or are replaced by some sort of loop. In any event some form of attachment is still present as opposed to those mounts under miscellaneous enamelled objects below which have 'no visible means of support'. On the whole shapes are either circular, often slightly domed, or rectangular. Function is probably as varied as the type.

Over fifty objects belong to this group. The enamel is found on the upper surface in a small number of cells which are circular or geometric in shape. Over half are enamelled in a simple technique of one colour to the cell. Blue and red are the colours most frequently met usually together though blue with green, yellow and white respectively were noted as well as red with yellow; there are two cases of red, blue and green. A fairly large amount of the enamel is now missing or decayed as on a rectangular mount with two studs from Brough now in the British Museum (reg. no. 74.12-28.38). Few of this class have a complex enamel and these are equally divided between alternately coloured pieces and grounds with spots inserted. An example of the latter was found at Chesters (Budge, 1907, p. 381, no. 1134). Again a large number, eleven, have millefiori enamel as on the two cow-hide shaped mounts from Leicester (Jewry Wall Museum, reg. no. 362.1951/45) and near Eden Hall in Cumberland (Victoria and Albert Museum, reg. no. 4898-1901)[71].

The majority of these mounts have been discovered in the north often on military sites such as Chesters or Corbridge. The rest of the country is evenly represented by the remainder. Few are satisfactorily dated. The mount from Eden Hall (q.v.) was found with coins ranging in date from A.D. 117-161. A pair of flower-shaped mounts with unusually long studs comes from an early second century context at Saint Albans (Frere, 1972, p. 126, no. 98). It is unlikely that all the rest belong to this century alone.

Belt-plates (Figure 9A)

Belt or belt-buckle plates are a numerous class of relic from Roman Britain. There are almost forty known, making up three groups. The first two are roughly equal in number and constitute the bulk of the type. Group 1 (figure 9Ai) consists of a narrow rectangular plate with two or more studs on the underside and with a buckle at the end as on that found at Caerleon and in the site museum there. Generally though only the plate survives and in a few cases the buckle alone. The second group consists of a less narrow open rectangle usually with pelta-shaped or wavy ends and with a separate bar inserted along the middle of the centre. A complete specimen comes from South Shields (Bruce, 1885, pp. 261-262) but again the bar is usually missing or else is found on its own. The third small group consists of a solid rectangle.

On the long narrow type - group 1 - enamel is found on most of the upper surface. It is simply applied and is generally red and blue where it survives or has not decayed. There is the rare use of yellow and possibly green. The design of the cells consists of either small geometric shapes (triangles, lozenges, crescents) enamelled in red and blue as on one in the site museum at Corbridge or else there are three long cells of laurel leaves in the centre and vandykes in the side cells as on another plate in the Caerleon site museum. Where the buckle piece survives it often has triangles filled with enamel on the upper surface.

The second type - group 2 - has enamel on the four sides of the open rectangle (figure 9Aii). Where this survives it is usually a millefiori enamel as on three specimens in the Corbridge site museum or else (e.g. once at Corbridge again or on one from Caerwent in the Museum and Art Gallery, Newport) there is a yellow enamel ground sprinkled with brown flecks - a rare technique previously seen on one of the seal boxes. Where the ends of the plates are extant they generally show simple enamel in red and blue. The bars normally consist of a series of three or four ovals - each a separate cell - all blue or else blue and red.

The best example of group 3 was recovered from Newstead (Curle, 1911, pp. 329-330, 332, pl. LXXXIX, 25) (Figure 9Aiii); this plate is entirely covered on its upper surface with enamel in both millefiori and the brown speckled yellow type just mentioned. One from Caerleon also has both techniques used together while a third from the same site (both in site museum) has millefiori used alone.

Overall the majority come from sites in the north and south-west of the country and not surprisingly almost always military sites as the four from South Shields, the five from Brough, eight from Caerleon and eight more from Corbridge. There is no obvious distinction between the three classes in this respect. As so often not many are satisfactorily dated. A group 2 specimen comes from a mid-fourth century floor in the commandant's house at Caernarvon (Wheeler, 1928, p. 139, fig. 61,10) and another from the fort at Ravenglass was found in a late fourth century context (Potter, 1979, p. 69, no. 13, fig. 26,13). Three of the group 1 type from Caerleon are dated to the second century. The superb group 3 plate from Newstead (q.v.) is probably Antonine while the long rectangular one from Richborough seems to have been found in a third century context (Bushe-Fox, 1949, p. 123, no. 73). One of the group 2 type from Wroxeter is included in a list of objects associated with the mid-first century military phase (Webster, 1958, p. 96, no. 258, fig. 8). We therefore have a very early and a very late group 2 and several second century specimens of groups 1 and 3.

Not a great deal is said on the origin of these objects though the consensus would appear to suggest a British manufacture for group 1 and a Continental one for group 2.

Finger-rings (Figure 8B)

Finger-rings constitute such a large and varied class of object that it is not proposed to try to deal with the enamelled examples in any detail but rather to note that this type of decoration is occasionally found on this class of jewellery (Figure 8B). It is best to note only those which have the enamel preserved or else where it is reasonably certain to have existed since empty settings on finger-rings are much more likely to have contained stones or intaglios and where a paste-like substance remains it is as likely to be the paste or glue to hold the latter in place as to be a decayed or weathered enamel. The number of enamelled finger-rings may therefore be much greater than appears.

Over thirty finger-rings decorated with enamel were noted. Four of these form a group of their own and are dealt with at the end. In all cases the enamel is found in the bezel and only rarely on the shoulders as well. There is great variety in the size of the bezels which are mostly circular, though some are oval in shape. For the most part there is only one cell. Blue seems to be the most common colour but red, green and white are also applied, in the majority of cases as a simple enamel. A good example of this simple enamel in blue is found on the ring from Newstead (Curle, 1911, p. 333). There are about half a dozen specimens with a more complex enamel, the most common being a blue ground with a few white spots inserted as on one of those from Traprain Law (Cree, 1923, p. 210, fig. 20,2). About a third of these finger-rings possess northern find-spots while the rest have what might be termed an expanded south-eastern distribution since few can really be said to have been recovered in the south-west. Dates appear to be rather early or rather late. One from Brough-on-Humber was found in a layer dated c.A.D. 125 (Wacher, 1969, p. 87, no. 7) while on the other hand an example from Richborough is probably late third century in date (Bushe-Fox, 1949, p. 128, no. 104) and a second from Saint Albans was found in a late fourth century deposit (Kenyon, 1934, p. 259, no. 7).

From Scotland there is a small group of large finger-rings which are discussed together (Simpson, 1970, pp. 105-108). The most recent find from Tarnaverie has red, yellow and a missing colour - blue is suggested - in separate cells forming a triquetral pattern (ibid. p. 105, fig. 1a (Figure 8Bc). Another from Lairg has a more simple triskele design but the enamel is missing (ibid. p. 106, fig. 1c)[72]. A third from Tentsmuir has a somewhat simpler design in yellow and a now decayed greenish colour (Scott, 1922, p. 109, fig. 1). Possibly also belonging to this group is one from the Culbin Sands which however seems to have suffered in a fire (Callander, 1933 pp. 32-33, fig. 8,1). Simpson (op.cit.) suggests an early date for manufacture, probably in Scotland.

Miscellaneous (Figure 9C)

There are over seventy objects which do not fit any of the previous groups. Some constitute small groups of their own such as the casket-handles while others are more individual such as some of the mounts. Many are unimportant, others are well known such as the Thames plaque. As far as possible they are dealt with in groups.

There is a small number of animals in the round, mostly cocks or hens. Some have a few settings which may not necessarily have contained enamel but on the example from Corbridge (site museum) there is simple enamel in red and blue on the crest, eyes and in squares on what remains of the breast. A religious function may explain some of these pieces.

Two enamelled bracelets are known, the more famous one from Southacre (Mottram, 1966, pp. 259-262) with simple red and blue and a second with green enamel from Baldock from a mid-second century burial (Westell, 1931, p. 263)[73].

What appears to be a candlestick was found at Bartlow (University Museum of Archaeology and Ethnology, Cambridge, reg. no. 48.1165) and is enamelled seemingly in red and blue. Another was found at Corbridge (site museum, reg. no. P325) and is enamelled also in red and blue and displays the laurel leaves, vandykes and toothing seen on some of the vessels. A third in the Victoria and Albert Museum (reg. no. 393-1872) has more geometric cells in red, blue and yellow[74].

There exists a small group of casket-handles (figure 9Ciii) such as that from Chesters with a complex enamel of a blue ground with yellow spots in addition to simple red (Budge, 1907, p. 385, no. 1397). The Lydney Park specimen comes from a layer yielding late third to fourth century coins (Wheeler, 1932, p. 86, no. 96, fig. 20) while two come from the Walbrook (Guildhall Museum, 1960, p. 8, nos. 1 & 2).

There are three short tubes with a band of lozenges on the outer surfaces enamelled in one case in red and blue (Figure 9Cii). This comes from Kirkby Thore (Budge, 1907, p. 406, no. 134) while the other two were found at Silchester (Reading Museum) and the West Park, Rockbourne, villa (Hewitt, 1968, pp. 30, 36 pl. Xb). They seem to be knife or strigel handles. Another slightly different handle with green and blue enamelled lozenges was found in an Antonine pit in Chichester (Down and Rule, 1971, p. 45, no. 8, fig. 3:16,8). A possible knife handle from Farley Heath has one panel of millefiori enamel (British Museum, reg. no. 53.4-19.121).

Another group consists of twenty-five various shaped mounts many of which do not call for any particular notice. There is however one type with nine specimens consisting of a long narrow rectangle, with slight sides turned down making it thus somewhat 'hollow'; on the upper surface of this is one long cell with a design of linked hearts (Figure 9Ci). Much of the enamel is decayed or gone but the ground seems to have been blue and within the hearts either red or green. These are enamelled in a simple technique. A good example comes from Wroxeter (Bushe-Fox, 1916, p. 26, no. 17, pl. XVI) and another from Richborough comes from a layer dated to c.A.D. 85 (Bushe-Fox, 1949, p. 133, no. 152, pl. XL). The rest are undated and distribution lies fairly evenly over the country. Mention may be made of two somewhat similar mounts except that these have a knob at one end and possess good Continental parallels (Figure 9Ciii). One comes from Campsfield (Hunter and Kirk, 1952, p. 61, no. 16, fig. 26,12) and the other from Shakenoak (Brodribb et.al. 1968, p. 86, no. 11, fig. 29). Both have millefiori enamel as do five other miscellaneous mounts.

Finally there are about a dozen miscellaneous objects some of which call for comment. The bronze plaque from the Thames is perhaps the best known and important because it is unfinished i.e.

the edges are untrimmed and possibly the enamel was never completed (Franks, 1870, p. 516; Brailsford, 1964, p. 56, no. 3, pl. XXI). In the Reading Museum, from Silchester, are two lids both simply enamelled in blue, green and red and with some signs of toothing to the edges. The latter is seen too on an incomplete plaque also from Silchester and believed to be part of a composite hexagonal vessel (Jope, 1955, p. 43, fig. 6).

CHAPTER 5

POST-ROMAN ENAMELWORK

The purpose of this chapter is to examine briefly the enamelwork from Britain in the few centuries after the departure of the Romans. That is to say up to about A.D. 700. There are few enamelled objects from later Anglo-Saxon times and these are not relevant to the present study[75]. It is intended to show what enamels there are from this period, what techniques are used and what continuity, if any, exists with the Romano-British industry. No detailed account is to be attempted and discussion of the types is avoided. It is best to avoid such controversial areas as the hanging bowl problem, the typology of penannular brooches and Irish enamel-working. These questions cannot form part of the present study though indeed solutions to them would be useful.

Compared with previous chapters the amount of material to be dealt with is not large. There are just over one hundred objects of rather varied types though approximately two thirds are parts of hanging bowls. It should be noted that much of the discussion on metalwork of this period - and here much more has been written on some of the types than on those of the preceding era - includes Irish finds, e.g. the penannular brooches (Kilbride-Jones, 1937, pp. 379-455), or brings Irish manufacture into the picture (e.g. Henry, 1936, pp. 209-246). This chapter is restricted to objects found in Britain and refrains from saying anything on the thorny problem of the origin of the hanging bowls. It deals with the smaller groups of objects first i.e. the brooches, pins, the Norries Law and Saint Ninian's Isle finds, the necklaces, the Sutton Hoo enamelwork and the miscellaneous pieces. They are treated in much the same way as the previous Romano-British objects.

We may begin with the few more Anglo-Saxon objects as opposed to the native British ones. Those bearing enamel include some cruciform, large square-headed and saucer brooches. In all there is hardly a dozen specimens. From Ixworth comes a pair of cruciform brooches with a spot and circle on the head and upper end of the foot; in the former there is no enamel remaining and in the latter only traces of decayed enamel. Both brooches are in the British Museum (Hills, 1871, p. 299). It is suggested that the two circular depressions on a late cruciform brooch from the Little Wilbraham cemetery are intended for enamel (Fox, 1923, p. 262) and traces of red enamel are said to be present in a spot on the bow and in two oval cells on the foot of another specimen found at Lakenheath (Jessup, 1950, p. 105). Several large gilt square-headed brooches from graves 21, 32 and 39 of the cemetery discovered at Linton "bore red or yellow plaques of enamel or paste" (Fox, 1923, p. 260) and two similar finds again from the Little Wilbraham cemetery "have small plaques of blue or red enamel" (<u>ibid</u>.p. 62). These brooches are however described elsewhere as having only red enamel (Leeds, 1949, pp. 57, 60, 69-70). Possibly the enamels are in fact decayed.

There appears to be one example of an applied saucer brooch with a spot of red enamel in the centre instead of the more usual amber, glass or garnet. This was found at Faversham and is now in the British Museum (Jessup, 1950, p. 62; British Museum, 1923, pp. 35, 48). There are then on a small number of Anglo-Saxon brooches, mainly of the sixth century, a few rather simple and small cells containing enamel. Some has decayed but red seems the most common colour with possibly some yellow and blue as well. The technique is always simple.

Despite the large number of penannular brooches with enamel decoration few of these come from Britain (Kilbride-Jones, 1937, pp. 379-455; Fowler, 1963, pp. 99-120, lists pp. 135-150)[76]. There is a type F1 example from the uppermost level at Traprain Law; it possesses a lozenge-shaped cell with yellow enamel on each terminal (Burley, 1955, p. 164, no. 89). A type H/F with enamel, as always with penannular brooches on the terminals, was found on the Scilly Isles (Fowler, <u>op. cit</u>. p. 143) and from Walls on Shetland, comes a type H3 with rounded terminals bearing spots of red enamel (<u>ibid</u>. p. 144). It is interesting that on this the enamel can be seen from either side i.e. there is no base for it. Finally there is a lead terminal of a F2 type pennannular brooch from Dinas Powys (Alcock, 1963, pp. 120-2).

There are somewhat over a dozen pins which appear to be intended for enamel decoration. The types are discussed by Fowler (1963, pp. 120-132) where reference to earlier work is made. Two types are of interest to the present study, the type E pin and the hand pin (including proto-hand pins)[77]. On the former the enamel is found on the thickened head of the object. The pin from Cassington has a pelta based cell with decayed enamel while a different more elaborate design occupies the head and upper stem of the specimen from Cirencester - here the dot and circle contain red enamel as well as the majority of the triangles, which otherwise are now empty (Brown, 1976, p. 21). It is claimed that the hollow in the head of the somewhat earlier type E pin from Newstead has red enamel and suggested that three others also with a hollow or dimple in the head were likewise treated (Fowler, <u>op. cit</u>. p. 121); this may have been so.

The hand pin has a relatively large head consisting of a flat projecting semi-circle along the top of which project in turn three to six 'fingers'. The front of the head bears enamel as well as (in most cases) the fingers. There is usually some sort of scroll or spiral design left reserved in the cell on the head. About a dozen are known from Britain, most with provenances in the north. Both silver and bronze specimens have been found. One of the bronze proto-hand pins (minus the fingers and with a complete circle as the head) has red enamel on the lower portion of the head. This was found at Halton Chesters in a late fourth century context (<u>ibid</u>. pp. 123, 126). A second proto-hand pin, but in silver, with a simple red enamel, forms part of a hoard from Oldcroft (parish of Lydney, Gloucestershire) deposited on the coin evidence between 354 and 359 (Johns, 1974, pp. 295-7)[78]. Among the best known of the bronze hand pins from Scotland is the example discovered at Urquhart with red enamel in the head but with empty settings in the fingers (Morrison, 1873, p. 357). Others have been recorded from

Pabbay, the Culbin Sands, Birsay and Kirbister (Proceedings Society Antiquaries Scotland, 1901, p. 278 and 1908, pp. 6-9 and 1930, p. 13). From Gaulcross comes a silver hand pin with now decayed enamel in the head (National Museum of Antiquities of Scotland, reg. no. L1962, 128) while on another of the same metal found on Freswick Links there is blue and red enamel - the latter decaying to green in places (Proceedings Society Antiquaries Scotland, 1947, p. 196). No less than three silver hand pins with traces of only red enamel in the heads and fingers were recovered with the Norries Law treasure (see below). Thus there are several examples from Scotland of bronze and silver hand pins with simple enamel, usually red only, on the heads and fingers[79]. The contrast may be made with the larger number of hand pins from Ireland where the specimens are of bronze and often include millefiori decoration.

Two hoards from Scotland contain enamelled material relevant here. The first was discovered at Norries Law about 1817 (for the best account see Anderson, 1881, pp. 34-7). It is housed in the National Museum of Antiquities of Scotland and includes five enamelled silver objects, the three hand pins and two leaf-shaped plaques (reg. nos. FC30-34 respectively; Durham, 1864, p. 8, pl. I). Two of the former are large (16+ cms.) and the third incomplete one small by comparison - all have three fingers. Much of the enamel is decayed or missing but there are traces of red which would appear to be the only colour applied. On the plaques the cells take the form of 'Pictish symbols' (Thomas, 1963, p. 45) and now have their probable lost enamel replaced with red wax[80]. The date of deposition of this hoard is disputed between the sixth and seventh century (Fowler, 1963, pp. 127-128). There is one instance of enamel in the Saint Ninian's Isle treasure (Small et al. 1973, pp. 54-55). The silver bowl, number 6, has an internal mount with a triangular cell of red enamel though there is the suggestion that the mount might not be original. The hoard itself is believed to have been deposited in the eight century (Wilson, 1964, p. 60).

Mention may also be made of a small but well known group of massive silver necklaces (chains) also from Scotland and connected with the Picts. The terminals do not always survive but in two cases where they do the enamel cells again take the form of Pictish symbols. On the example from Parkhill the enamel is missing (National Museum of Antiquities of Scotland, reg. no. FC147) and on a second found at Whiteclaugh (ibid. reg. no. FC150) the red wax in the terminals replaces the lost red enamel. It seems likely that red was the original colour.

Leaving Scotland for the present there are relevant pieces in the Sutton Hoo find[81]. The most spectacular enamelwork is to be found on the large hanging bowl which bears this type of decoration on the three circular escutcheons, three square plaques set between these, the interior print from which there stands up a fish in the round and an exterior print (Bruce-Mitford, 1966, pl. 10 and 1979, pp. 40-41, fig. 26). On the escutcheons the outer cell with the reserved scrolls contains red alone though in places this has decayed to a green colour; the narrow middle cell contains translucent blue enamel with small squares of millefiori glass floating in it; the centre cell is again red with various pieces of millefiori floating crowdedly within it (*ibid*. pl. 4). The outer cell of the square plaques again has red enamel but in each corner is a separate leaf-shaped cell containing green enamel which appears to have had small pieces of another colour (?red) inserted into the surface; the square centre cell again has many pieces of millefiori floating in a red ground (*ibid*. fig. 28). The interior print bears plain red enamel in the outer narrow band while the inner band contains blue enamel with millefiori once more floating in it. The body of the fish is covered with now empty spots, 2 mms. in diameter, once possibly holding enamel. The outer cell of the exterior print once again has red enamel in a scroll pattern but possesses only one cell within having a red ground, a medium translucent blue pelta-shaped pattern and various pieces of millefiori floating around the latter.

The details of the other two hanging bowls are less clear. Only the interior and exterior prints of hanging bowl no. 2 bear enamel (*ibid*. fig. 62). The style is different from the designs on the large hanging bowl. It consists of a rather heavy Durrow type animal with four heads emerging from a central spot - a "zoomorphic swastika design". There is not much room left for the enamel cells which seem to be simply filled with red enamel. Hanging bowl no. 3 has three escutcheons with four thin peltae only reserved (*ibid*. fig. 63). The simple enamel is red but each escutcheon possesses a central millefiori inset (Bruce-Mitford, 1975, p. 441, nos. 111 and 112).

Some of the pieces among the gold jewellery from Sutton Hoo are also of interest, namely the purse-lid and the pairs of shoulder-clasps and pyramids. These all bear millefiori insets. Detailed examination of this inlay shows that it consists of slices of millefiori glass treated in the same way as the garnet inlay (Bruce-Mitford, 1978, pp. 582-597). This does not therefore come within the scope of enamelwork but is worth noting as the use of millefiori glass in a somewhat unusual technique.

There is no need here to discuss the dating and background of the deposition of the Sutton Hoo objects. Recent work on the coins suggests 625 as the most likely date for their assembly though care needs to be taken in the interpretation of this (Bruce-Mitford, 1979, pp. 82-94). In combination with the wear on the hanging bowls, particularly shown in the patching of the large bowl (*ibid*. p. 41, fig. 29), this puts the date of manufacture and enamelling of these three bowls back to the previous century.

Before moving on to the hanging bowls as a group there are a few more miscellaneous enamelled objects from this period worth noting. A bronze finger ring with four triangular cells, alternately red and yellow, on the bazel was found during the excavations of the monastic site at Whitby (Peers and Radford, 1943, p. 58, no. 49) and a second one also possibly set with blue enamel (or 'paste'?) comes from grave 100 at the Long Wittenham cemetery (Akerman, 1860, pp. 346, 352). Although the mount from Liddington has been listed as belonging to a hanging bowl (Fowler, 1968, p. 308) this is hardly the case and it should perhaps merely be termed a 'mount'. The enamelled pattern is basically a pelta though the enamel is now decayed. A mount from Colchester with some sort of animal pattern on it was said to bear traces of enamel but the object is now lost (Hull, 1958, p. 92). From the site of an Anglo-Saxon cemetery at Fetcham a small circular mount bearing red enamel in a simple pattern of petals and curved triangles was recovered in the 1930's (Cotton, 1933, pp. 48-49). Mention might be made too of a few examples of Romano-British enamelwork which survived into this later period. Among these are the head stud brooch found in grave 2 of the Anglo-Saxon cemetery in the grounds of Girton College, Cambridge (Fox, 1923, p. 247), the mount with millefiori enamel from Sarre (Jessup, 1950, p. 119) and the cow-hide shaped mount, also with millefiori enamel, from York (q.v.).

This leaves the hanging bowls. It is proposed only to look at the enamel aspect of this group of objects. So much has been written about the contexts, typology, origins, functions and dates of hanging bowls yet the basic requirement of a corpus is still sadly lacking. Henry (1936, pp. 209-246) is probably still the most useful from the point of view of the actual objects though a more up-to-date list with useful comments has been contributed by Fowler (1968, pp. 287-310) where reference will be found to other works[82]. A list of those hanging bowls bearing enamel, noting only the present colour or state of the enamel and the technique employed, is given in Appendix B.

To date over sixty hanging bowls bearing enamel decoration have been recorded. These are represented either by finds of single escutcheons or complete bowls bearing escutcheons and other mounts. The applied escutcheons, normally three in number, are the usual places to find the enamel but enamelled discs, known as prints, also occur in the centre of the base, both internally and externally. In addition, on some bowls there are applied, on the exterior surface, plaques, sometimes in the shape of animals, and bands also, bearing enamel. A good example of the survival of a single escutcheon is that from Chalton (Addyman and Leigh, 1973, pp. 19-20) while Lowbury Hill yielded an example of a bowl with its three escutcheons and two prints still attached (Atkinson, 1916, pp. 21-22). The addition of square plaques is to be seen on the large hanging bowl from Sutton Hoo (where note also the free-standing fish; Bruce-Mitford, 1966, pl. 10) and bands as well as zoomorphic plaques are applied to the Hildersham, (Lethbridge, 1952, pp. 44-47) and Lullingstone bowls (Kendrick, 1932, pp. 171-173).

Mention may also be made of a few more recent finds. The Chalton escutcheon (q.v.) is one of these; its enamel has now decayed to a greenish colour and the design is based on the Durrow spiral pattern (Fowler, 1968, group IV). Another enamelled escutcheon with a bird-headed triskele pattern is recorded from Wilmington, Sussex, (Wilson, 1969, p. 241) and another escutcheon or print, again enamelled and said to be similar to the Chalton specimen, was found during the excavation of the Anglo-Saxon cemetery at Bevis's Grave, Bedhampton, Hampshire (Webster, 1976, p. 161)[83]. Also there is the lead model for an escutcheon, of an enamel-bearing type, from the Brough of Birsay (Curle, 1974, pp. 301-307)[84]. Red, yellow and blue are the colours of the enamels used on the hanging bowls. However the colours given in publications may not always be correct - unless decay has occurred since the original publication. For instance on one group of escutcheons illustrated (Leeds, 1933, pl. III) the following is found: no. 4 from Barrington (B) shown as all yellow, ought really to be described now as decayed, and likewise no. 5 from Greenwich shown as being all red. Again one would be inclined to describe the enamel on the Benty Grange escutcheons as decayed rather than yellow (Bruce-Mitford, 1974, p. 250) though this is preferable to yellow on a red ground (Henry, 1936, p. 236). It is obvious care must be taken in describing enamels and though Leeds' assumptions for instance may be correct it seems preferable to err on the side of caution and list such cases as 'decayed'. Green does not appear to have been a colour intentionally used (except in the one case of the corner cells of the square plaques on the large hanging bowl from Sutton Hoo, q.v.) and now greenish colours are the result of decay as seen on the Chalton escutcheon.

Red is the most common colour found, on over a third of the finds, both on its own as on the Lullingstone bowl or in combination with yellow or blue. Red and yellow are to be seen on more than half a dozen specimens such as on the Hitchin escutcheon (Victoria and Albert Museum, reg. no. 163, 1923). Yellow alone is probable on the Camerton escutcheons (Leeds, 1933, pp. 150-151) as well as the Barrington B escutcheon and the Benty Grange escutcheons (both q.v.) There is also the combination of red, yellow and blue on the escutcheon from 'near Oxford' (Leeds, 1933, pl. III, 7) but otherwise blue does not appear to have been frequently used (except see the hanging bowl from Sutton Hoo above). About a quarter of the enamels examined on these objects would need to be described as 'decayed' but in only a few cases is it gone entirely e.g. that from Eastry (Henry, 1936, p. 228, pl. 25,3). So red enamel is by far the most frequently used followed by yellow and often with the two combined. Otherwise only blue is used but rarely and green in one minor instance.

The overwhelming majority of these enamels are simple with one colour to a cell and often two colours to an object except for one case of three colours. There are rare cases of what might be described as complex enamels in so far as on the 'Hitchin' and 'near Oxford' escutcheons the red and yellow enamels meet in places without a septum between them; the point is well illustrated by Leeds (1933, pl. III, 6 & 7). In addition millefiori occurs in five cases. These are the Barlaston escutcheons (Kendrick, 1932, pp. 173-174, fig. 5 & pl. 5), the Northumberland escutcheons (Smith, 1908, p. 73, fig. 2 & 3), the Scunthorpe escutcheon (Kendrick, 1941a, pp. 236-238, pls. LII & LIII), the escutcheons, prints and plaques on the large hanging bowl from Sutton Hoo (q.v.) and the lost hanging bowl from the River Witham (Kendrick, 1941, pp. 161-162). On most of these the millefiori is clearly floating in a ground of red enamel. There is the suggestion (Henry, 1936, p. 237), which should perhaps not be entirely disregarded, that the spots on the 'near Oxford' escutcheon (q.v.) consisted of millefiori too[85].

The greater number of the escutcheons are of bronze and they are frequently tinned (see Appendix B). Only the print of the Saint Ninian's Isle bowl is silver and since the lost River Witham bowl was also of that metal so probably were the escutcheons. Most of the escutcheons and prints are circular, generally being about 4 cms. in diameter though there is a small number of about half the size; otherwise they are heater- or bird-shaped, square or zoomorphic.

For the most part the enamel cells are delineated by the designs or patterns, mostly based on scrolls and spirals. These can be gauged from the various discussions (e.g. Fowler, 1968, table 1, pp. 296-297; Longley, 1975, figs. passim, pp. 15-30). Here is not the place to examine them yet again nor indeed to try to date them. Dates of deposition can be assigned to such finds as those from Sutton Hoo or Whitby Monastery, though age when buried is an important factor. Yet few are likely to have been made after the seventh century and some must be as early as the fifth. Find-spots are mostly south-eastern in Anglo-Saxon contexts and on origin suffice to say that few would now probably agree with Henry's contention that they are all Irish.

Surveyed as a whole, there is not really very much enamelwork in the post-Roman period and it becomes very scrappy without the hanging bowls. Even with the latter most of the enamel is simple in one or two basic colours; the millefiori enamels are in fact few and hardly of the quality of the preceding period. It is generally recognised nowadays that in the field of metal-working (e.g. Clarke, 1951, p. 199) there was continuity before and after the Roman departure and if, as appears so, there were late Romano-British enamels being manufactured then the post-Roman examples can be seen as the final stage of this industry. The problem of a sudden re-birth of enamel-working in the early Anglo-Saxon period in fact does not exist either in volume or lack of immediate predecessors.

PART 2 THE TECHNIQUES

INTRODUCTION

Now that the available evidence has been surveyed what can be said about it? The most important part of the complete object is of course the actual enamel or vitreous substance which is dealt with in Chapter 6. There the colours applied are discussed and their analyses examined. Such aspects as raw materials and decay are included. Chapter 7 covers the second important part of enamelwork, the metal base. This looks at which metals are suitable for enamelling and examines each in turn as regards its use in the period under review. Analyses of the metals of various objects are examined - as well as the cells, their size, shape, mode of production and the various means employed to help keep the enamel in place. The following chapter (8) looks at the process of joining the vitreous substance to the metal base. The actual problem is discussed followed by a survey of the tools needed to carry this out and a brief look at what can go wrong. Then follows a suggested arrangement for grouping these early enamels on the mode of application: simple, complex and millefiori rather than the all-embracing champleve Millefiori enamel is examined in some depth. In Chapter 9 comes the problem of actual sites where the work was carried out. The well known sites of the Continent and Ireland are first discussed and from these the types of evidence available are listed and the criteria then applied to a list of sites in Britain which are divided into probable, possible and doubtful manufacturing locations.

CHAPTER 6

THE VITREOUS SUBSTANCE

The actual enamel or vitreous substance is examined first. We look at what colours are used, which appear on objects of each period, the type of object they adorn, in which combinations they are employed and whether they are opaque or translucent. It should be noted however that here we are dealing strictly with the enamel and that the millefiori glass, in whatever technique it appears, is discussed in chapter 8. An attempt is made to discover just what the actual enamel is made of by examining the various analyses of the substance available in print. From these it is hoped to define the basic vitreous substance, the colourants added, the question of opacity and to compare these with the glass of the period. Raw materials are briefly examined and the problem of weathering mentioned. The end result should be a more comprehensive and useful study of the basic material constituting an enamelled object of the period under review.

Before looking at the colours used it should be said that each individual colour shows a wide range of shades. Some indication of these is given where appropriate but no attempt is made to describe each colour rigidly or to give it a code according to colour schemes such as the Munsell Colour Chart or the chart published by the British Horticultural Society for flower shades. On the whole this would not appear to serve any useful function and in any event is liable to subjective observation as well as the decay of time. Unless it is reasonably obvious what the colour is, it is described as being decayed i.e. if it is said that six blue enamels were noted this means there were six blue enamels and not, as has often happened in the past, three of blue and three of a decayed colour which might once have been blue. Occasionally there are border-line cases as between blue and green and orange and yellow, and difficulties arise too with weathered enamels now seemingly green or white, and in the end these have to be subjectively assigned to a group.

As already seen an opaque red (often described as 'sealing wax red') enamel is the most common on the Iron Age objects where most frequently it is found alone as in the Stanwick and Santon hoards. Yellow and blue occur in much fewer instances and then only as spots or in small angular cells. Neither is used on its own and it is discussed later whether the blue is an enamel at all or merely a glass inset. Red and blue, as on the two mounts from Westhall, is a combination slightly more common than red and yellow, exemplified by the massive armlets from Scotland. Yellow and blue only are unusual together but there are several cases of the three colours, red, yellow and blue, combined e.g. on the two terrets from Saham Toney. However these three colours are the only ones that can be said to have been applied to Iron Age objects and shades of green or white should be regarded as examples of decay

The enamelwork of the Romano-British period is marked by a greater variety in the enamel colours. In addition to red, blue and yellow there is now green, orange, black and white. Surprisingly blue becomes the most common colour, having been recorded on over five hundred objects. There is much variation in the intensity of this shade ranging from a dark or deep blue through medium and light blue to a rather watery blue which almost verges on white at times or on other occasions is near a light green or turquoise. Red survives as a certain colour on somewhat less than four hundred objects. On the whole it is not such a brilliant red as that of the Iron Age and has fewer degrees of shade though it sometimes merges with orange. Possibly if the original colours of the decayed enamels were known this imbalance in numbers between red and blue might be adjusted - though there is no indication that this would necessarily be so. Yellow comes a long way behind in popularity adorning about a hundred pieces and ranging from a pale yellow to a deeper sulphur colour; again it comes close to orange at times. There is an almost equal number of green enamels, that is to say what appear to be intentional greens as opposed to other enamels which have decayed to a greenish hue; the genuine green tends to be light or medium, very rarely dark. Orange does seem to be present on a few dozen specimens in its own right rather than as a yellow or red which has not turned out to be the correct shade. Black is used in a small number of cases especially as spots or circles in the complex technique. White too is found, often a rather ivory white, again frequently as spots; in many cases it is difficult to be sure whether this colour is an intentional white or a white produced by the decay of another colour, but there is no doubt that there are instances of the former. There is the odd case of genuine brown but more often it might be that such a colour is really a decayed red. This applies more often to descriptions in older publications and decay too may account for any 'grey' listed. It seems reasonable to accept that there are seven colour found on Romano-British enamelwork and that blue and red are by far the most common.

Though there are cases of only one colour being found on these objects it is much more usual to have two colours and even three during the Romano-British period. Blue and red is by far the most common combination being found for instance on four times as many objects as the next most popular pair blue and yellow. Red also occurs with yellow as well as green, black, white and orange. Other than red, blue is combined with green and orange. Red, blue and yellow and red, blue and white are the main triple combinations though occasionally other groupings of the various colours are employed by the craftsman. This applies to those simple enamels where only one colour is applied to a cell. With the more complex enamels there appears to be a greater variety of colour schemes. Where there are two different colours in the one cell red and blue and red and white are often found as well as blue with white or less often with yellow, green or orange. In a similar fashion red, blue and yellow or blue, white and green or red, blue and white have been noted. In those cases where the complex enamel consists of a background colour with spots of a second inserted the former is most commonly found to be blue or red, less often orange, yellow or green while the spots are one of red, yellow, black, white or blue.

Blue and red are the most frequent colours on the head stud (group Q) and trumpet headed (group R) brooches with yellow appearing less often. The group S brooches too have mainly blue and red but here the third colour used is green. On the dragonesque variety yellow is almost as common as blue and red. Only the latter two are employed on the domed plate brooches. Blue occurs on the greater number of flat disc brooches followed closely by red and on a lesser number green, with small amounts of black (see especially type 11), yellow and white. The sandal shaped (animal) plate brooches provide a good example of complex colours with generally a blue ground with white or yellow spots. The vessels are often decorated with a combination of red, blue and green while on the seal boxes, miniature stools and button-and-loop fasteners the usual colours are blue, red and yellow in that order.

As noted in chapter 5 the number of colours used in the post-Roman period drops to three. Then red is by far the most usual, followed by yellow and in a few cases blue.

Although we deal later with the question of what causes opacity since it often arises in relation to enamelwork it should perhaps be noted now that this interest in opacifiers is due to the fact that the overwhelming majority of the enamels examined for this study are opaque. Despite the fact that nowadays translucent enamels are as common as opaque ones, very few of the former occur in the periods under review. The small number of translucent blue spots from the Iron Age are reasonably well known though these are glass rather than enamel (see below). Otherwise there are only about a dozen examples of translucency on Romano-British enamelled objects. These translucent enamels are generally blue such as on a group M brooch from Caerwent (Newport Museum), the lozenge-shaped seal box lid found at Corbridge (site museum) and the flat disc brooches recovered also at Corbridge (type 1 - site museum) and Richborough (type 8 - Bushe-Fox, 1949, p. 116, no. 45). It is possible that these are not intentionally translucent and could be due to lack of control over the constituents of these particular enamels.

Before looking at what the enamels are actually made of it is necessary to mention the possibility of the use of scrap glass or cullet, particularly as colouring matter. The secondary use of such material has for example been shown in the bangle from Traprain Law made from a Roman pillar-moulded bowl (Newton, 1971, p. 13).

We begin by looking at the Iron Age enamels in detail, especially the red enamel. Various analyses of this red enamel have appeared since the nineteenth century but the only complete ones are still those of the stray lump from Tara (Ball and Stokes, 1892, p. 280) and of various pieces from Bibracte (Bulliot, 1899, p. 146 and p. 35) though see below. A few later analyses deal only with the colourant for example the Trevone brooch (Dudley and Jope, 1965, pp. 18-23) or the brooch from Eastham (Stead, 1965, p. 65). More recently a group of thirty-three objects of the Iron Age bearing this opaque red enamel, or 'glass' as Hughes prefers to term it, was analysed to increase the amount of information (Hughes, 1972, pp. 98-107). The base glass was found to be of the alkali-lead-silica variety though only qualitative not quantitative analyses were carried out (*ibid.* p. 99). The colour, as was previously known, is due to numerous feathery crystals of cuprous oxide floating in the glass matrix and this form of copper is only obtained by heating in reducing conditions. The presence of lead is held to increase the solubility of copper and therefore because of their importance, the lead and copper contents were quantitively determined. The average composition proved to be 7 per cent cuprous oxide and 25 per cent lead oxide with there being more variation in the amount of the latter present in the red enamel (ibid. table 1, pp. 102-103). The make-up of this material is therefore reasonably well known and it was for this reason that only one Iron Age object, the terret from 'Suffolk', was included in the analyses carried out for the present survey (Bateson and Hedges, 1975). Lead oxide here amounts to 17 per cent and cuprous oxide to 6.4 per cent, both levels fitting in quite well with Hughes' ranges.

However it was mainly on account of the second colour, blue, that this 'Suffolk' terret was included. Much has been said on the blue spots of the Iron Age being in fact translucent blue glass insets held in place by red enamel but none had been closely examined before. When looked at through the microscope red was clearly seen beneath the translucent blue spots which were then analysed. In contrast to the true enamels lead was found to be absent and the blue colour, rather weak here, may be due to the trace of cobalt rather than the small amount of copper present (op.cit. pp. 178, 180 and 184). Similar translucent blue glass spots occur on the Bapchild terret (Leeds, 1933, pp. 40-41, 104) and the terret from Richborough (Bushe-Fox, 1949, p. 106): light again in the former but dark blue in the latter. The well known pair of glass spots on the Hod Hill stud are a translucent medium blue and provide an excellent example of this technique of insetting (Brailsford, 1962, p. 16). This might suggest blue enamel, as opposed to blue glass insets, does not exist but such is not the case. There are instances where the position is not clear as with the translucent blue squares on one of the terrets from Colchester (Colchester & Essex Museum, reg. no. 5233.26), the opaque spots on the Middlebie terret or the opaque petals adorning the Rise bit (Smith, 1925, p. 103, pl. VIII). However Leeds (1933, p. 104) thought that the light blue on the Saham Toney terret was enamel as also the blue on one of the quatrefoil harness mounts (ibid. p. 112). Likewise the blue in the heart-shaped cells on the Elmswell panel (Corder and Hawkes, 1940, pp. 338-356) and the light opaque blue on the Fayum terret (Smith, 1925, p. 87, fig. 81) both appear to be enamels. There are then several instances of the use of a light blue opaque enamel as well as spots of translucent light to dark blue glass held in place by red enamel. None of the former has however been analysed. Unfortunately there is a lack of knowledge concerning the make-up of the yellow enamel of the Iron Age.

Moving on to the Roman period there were until recently even fewer analyses available and in no case a fully quantitative one. Only about a dozen varied objects had had their enamel analysed with no overall plan and in all instances only qualitatively, mainly as regards colouring matter or opacifying agent. The main group of analyses available was that carried out on some of the British Museum's head stud brooches, though on these the enamel is often weathered (Painter and Sax, 1970, pp. 153-174). The reds were found to be due to copper and in two instances the opacifiers for now weathered enamels proved to be sodium antimonate and calcium antimonate respectively. A somewhat earlier examination of the enamels applied to the Bartlow vase showed copper to be responsible for the red and green colours but the blue results from the use of cobalt (Gage, 1836, p. 307).

Similar colourants are used for the red and pale blue enamels on the Southacre bracelet and the opacifier of the latter is again calcium antimonate (Mottram, 1968, p. 259). The blue/green enamel on the Wall brooch was also found to be due to copper while manganese seems to be the colourant for the brown enamel; in both cases lead is present (Werner and Organ, 1964, pp. 46-47). Lead, copper and silica were found to be present in the red enamel on the Carlungie brooch while the colourant for the yellow was shown to be lead antimonate and the blue was described merely as 'glass' (Wainwright, 1953, p. 68). The red on the dagger sheath from Hod Hill is the normal type consisting of cuprous oxide in a lead glass matrix (Brailsford, 1962, p. 5) while the opacifier of the yellow enamel on the second dagger sheath from Hod Hill was shown to be also the result of calcium antimonate (Richmond, 1968, pp. 137-138)[86].

It would appear from these results that copper was used extensively to colour red, green and blue though in two cases blue was produced by cobalt. Otherwise antimony was employed to produce yellow and manganese for brown. Sodium antimonate and calcium antimonate were both used as opacifiers. Finally lead was discovered to be present in several cases though in the remaining examples it was not necessarily sought. There is therefore a certain amount of information to be obtained from these analyses though they are rather haphazard and incomplete. It was thus decided to undertake a more comprehensive scientific study of a specially selected group of enamelled objects to see what exactly the enamels consist of and to try to discover any historical or geographical patterns and to examine any relationships to contemporary glass. Thirty enamels, on twenty-five brooches, were analysed for this purpose. These enamels form a representative selection of the colours in use during the Roman period while the brooches include both early and late examples of types manufactured either in Britain or imported hither from the Continent. The enamels sampled were chosen for their strong colour and lack of weathering or decay[87].

The constituents of the basic glass matrix (silica, soda, potassium, magnesium, calcium and aluminium) show a consistency in contrast to the rest of the data. The silica content was generally not measured directly but estimated by difference. Where more complete analysis allowed it was shown to average 55 per cent, though ranging between 35 and 70 per cent and therefore displaying a lower average amount than that in contemporary Roman glass where it falls into a smaller range of 60-68 per cent (Caley, 1962, pp. 64-65; Forbes, 1957, pp. 221-228). In the case of the flux, here sodium oxide, the estimated average content is 13 per cent while the other alkali oxide present, potassium, occurs in amounts generally less than .5 per cent and rarely rising to 1 per cent. This is consistent with soda rather than potash being the flux at this period and suggests that the presence of potassium is accidental. The calcium oxide content is much as might be expected, never exceeding 15 per cent nor dropping below 6 per cent and on average 9 per cent. Only selected specimens were examined for aluminium and magnesium. The amount of the former oxide varies between 1.6 per cent and 3.6 per cent while the magnesium on the whole is less than 5 per cent. In both cases the inclusion does not appear to have been intentional and could arise from impurities in the raw materials.

Another major ingredient is lead. Lead oxide averages 9 per cent and though there is much variation, ranges from 0 per cent up to 35 per cent in the samples analysed. Only the blue glass spots on the 'Suffolk' terret (q.v.) contain no lead. In contrast lead occurs rarely, and even then usually only as a trace element, in Roman glass (Caley, 1962, pp. 93-104) so that a major difference is established between glass and enamel of this period. Such a distinction might be useful in deciding whether a piece of raw material or waste is to be associated with glass-working or enamelling. Hughes has shown that lead oxide is present in the Iron Age red enamels where the typical level is 25 per cent. The Roman red enamels are therefore chemically similar to the Iron Age ones but display greater variation in the lead content, more akin to the other coloured enamels. Lead may be added for various reasons. It creates a softer, more brilliant, substance though one more prone to decay. More importantly it can lower the melting temperature or affect the coefficient of expansion or the high temperature solubility of copper or tin. It also plays its part as a colourant or opacifier.

Iron oxide is to be found in all examples varying between .7 per cent and 14 per cent. However in only two specimens is it found in significant quantities viz. 10 and 14 per cent in the black enamel on two brooches (op.cit. nos. 11 and 22). Otherwise iron constitutes less than 3 per cent of the ingredients and is lowest in the paler colours. In such cases it is most likely present as an impurity in the sand as most sands do contain iron oxide. Iron as an impurity and in the absence of any other colouring agent would produce a dirty green or brownish or even pale blue glass (Harden, 1956, pp. 312-314). It is therefore likely to have been a deliberate addition to produce black.

Cobalt was discovered in only nine of the samples analysed and in each case this is in a blue enamel. All the blue enamels analysed contain cobalt oxide. The amount however is small being only a trace in number (24)[88], .1 per cent in number (20) and consistently .2 per cent in the remaining seven specimens. Nevertheless cobalt is a strong colouring matter and these amounts would be sufficient to produce a strong colour and in fact in all these cases the shade is a deep blue (Taylor, 1977, p. 4). In the one example, (24), which has only a trace of cobalt the colour is weak, watery and translucent. Since the common raw materials of glass do not contain cobalt its presence can hardly be accidental.

Copper occurs in all the enamels though in no case does it exceed 9 per cent. It can be responsible for varying shades of blue, red and green (Harden, op.cit. p. 214). These however depend on its oxidisation state and co-ordination. In the nine examples of blue analysed the amount of copper present nowhere exceeds .8 per cent and therefore does not appear to have been added specifically as the colourant, especially in view of the presence of cobalt. It would seem that copper is responsible for the three greens examined as it is present (as CuO) in amounts of 1.9 per cent, 4.6 per cent and 5 per cent respectively. Red cuprous oxide (Cu_2O) is the colourant of numbers (17) and (24) where the percentages of 3.6 and 6.4 are near enough to Hughes' Iron Age average. The amount in numbers (1) and (2) - 1.4 per cent - fits less well but microscopic examination again showed a thick cloud of cuprous oxide crystals. Copper is also present

(Cu_2O) in quite high amounts in two orange enamels where it accounts for 5.5 per cent of (12) and 9 per cent of (23). It may be noted that there is significantly less copper in the white enamels where it might cause discolouration. If this is not deliberate, its presence may be explained as an impurity in the raw material or possibly a transfer of corrosion products from the metal base.

Tin oxide was found in most specimens ranging from a trace up to 5.5 per cent but on the whole is 2 per cent or less. Antimony is also present in most samples accounting for as much as 12 per cent of number (5) and overall being more abundant than tin. It appears from various published results that lead antimonate and calcium antimonate are common opacifiers and these may also be the colourants of the yellow and white enamels. Tin lead oxide and tin oxide could however perform the same functions. Harden (ibid.) suggests a suspension of tin oxide as producing an opaque white in Roman glass but in Forbes' (1957, pp. 221-228) analyses tin only occurs once - in a green mosaic cube (ibid. p. 224). The results for the white enamels showed little lead or tin but much antimony and suggest calcium antimonate was used to produce them; an X-ray diffraction study showed this to be the case with number (10) and the same method indicated that lead antimonate is responsible for the yellows of numbers (3) and (18). As the tin content is low in these samples calcium antimonate could be the main opacifying agent. Antimony is only absent in the translucent blue of (24) and is low in the reds which derive their opacity from the cuprous oxide crystals. There is much antimony in the blue enamels and calcium antimonate was suggested as the opacifier in the blue of the Southacre bracelet (Mottram, 1968, p. 259).

This leaves manganese and zinc. The former, with few exceptions, is less than 1 per cent and would appear to be an impurity in all cases either from the sand where it is common or even cullet where it may have been added as a de-colourant. There is however the case of the Wall brooch (q.v.) on which the brown was attributed to manganese though the level was not recorded. Zinc is present in a few samples but as the range is only between .2 per cent and .8 per cent it probably represents an unintentional addition, possibly introduced along with the copper.

Overall then the matrix of the Roman enamels examined consists of an alkali-lead-silica base, containing some not unexpected impurities, to which various colourants and opacifiers have been added. The amounts of the basic ingredients do not differ much from those of contemporary glass except for the presence of lead which is not a constituent of the latter. However there is much more variation in the amounts of lead, tin and antimony possibly suggesting these along with the colourants were added to a basic glass frit. Such a two-stage process was found to have been carried out at Wadi el-Natrum, the centre of one of Rome's main glassworks (Saleh, et.al. 1972, pp. 143-146). Although there is some variation, the red enamels may be taken as a continuation of the Iron Age opaque red enamel.

A second group of analyses has now been carried out on the enamels on a selection of brooches from the Romano-British site of Nornour (Biek et.al. 1980, pp. 50-75). The forty-five enamels examined occur on a variety of bow and plate brooches made both in Britain and on the Continent and dating mainly from A.D. 50 to 250. Eleven examples of red were analysed and found to contain less copper than the reds included in the earlier studies: a distinction is suggested between the brighter sealing-wax red, a duller red and a brick red in which copper in the colloidal state produces that colour and antimony the opacity. Cuprous oxide emerged as the colourant for the six orange enamels in which a variation in particle size may account for the orange rather then red colour. No green or yellow enamels were included in the study. The results confirm that calcium antimonate is the colouring agent in the ten white enamels though it is not clear what produced the black colours. All eleven blue enamels, which are less intense in colour than those examined by Bateson and Hedges, were found to owe their colour to cupric oxide and their opacity to calcium antimonate.

Colourants are perhaps one of the most interesting aspects of enamel or indeed glass studies. While modern text books speak in terms of adding .1 - 3 per cent of the colouring agent (Shaw, 1972, p. 155) it is surprising to find up to as much as 10 per cent present in, for example, the black of the Roman enamels. Most of the enamel colourants are similar to those used in early glass (Lucas, 1948, p. 219).

Sealing-wax red enamel was produced by cuprous oxide with the assistance of lead. Hughes has pointed out how difficult it is to produce a red enamel in the necessary reducing conditions <u>in situ</u> and has suggested it was emplaced, preformed, in a slightly molten state rather than as a frit or powder. Yet it will be seen below that the enamellers displayed much skill in the use of millefiori and this ability along with the common use of red enamel on Romano-British enamelwork may suggest they had overcome the problem. Reducing conditions can be achieved by fusing a frit under charcoal[89]. There are moreover less brilliant reds and in some of these the colour may be due to copper in the colloidal state. There is thus some variety in the colouring method of the red enamels.

Cuprous oxide, possibly because of a different particle size, is also responsible for some of the orange colours though other oranges appear to be due to a mixture of red and yellow where the yellow owes its colour to lead tin oxide. This mixture may have been of two enamels or merely of the two colourants added to a basic frit. Lead tin oxide yellow seems to occur only in orange enamels. Plain yellow enamels appear to owe their colour to lead antimonate.

Red and green enamels have a similar composition with the colour in the latter produced by copper in the oxidised state (cupric oxide). Furthermore the function of the lead here is to provide the necessary co-ordinating environment for the copper to produce the green colour, otherwise a blue would result. It is possible that the slow oxidisation of a red enamel (cuprous oxide) might produce a green colour. However no borderline case was observed and the colours of each are rich and homogeneous. Black seems to have been produced more simply through the addition of large amounts of iron and white has consistently been observed to be due to calcium antimonate, not tin.

Perhaps the most unexpected colourant is that for the deep blue where in all cases analysed it proved to be cobalt. Harden (1956, p. 314) listed copper or iron for this purpose in glass since cobalt had "not so far been discovered in glass throughout our period in the West". This absence of cobalt in blue glass is also shown in the analyses of Forbes (1957, p. 126) although Caley (1962, p. 102) and Gunther (1912, p. 108) give three instances. The blue on the Southacre bracelet was attributed to cobalt though copper appears to have caused the Wall brooch blue. All the blues analysed by Bateson and Hedges are quite intense (except no. 24) and therefore unlikely to be due to copper which would result in greenish blue because of the presence of lead. Copper was however found to account for the many paler blues encountered in the study of the Nornour enamels. A blue colour therefore is, depending on its intensity, due to either cobalt or copper.

Something should perhaps be said on the sources of these various elements. Little work has been carried out on this aspect of enamels though comments have been made relating to the basically similar process of glass-making. The main ingredient, silica, is perhaps the easiest raw material to obtain since it constitutes the bulk of sand. However sand is rarely just pure silica, although colour can give some guidance on the level of impurity, e.g. the fine white sand obtainable locally was probably used at the Romano-British glassworks at Wilderspool and is much preferable to a darker one containing more impurities. Iron is usually the chief of these and this is what gives the basic glass its normal greenish or brownish colour. Flint, sandstone, quartz (seemingly used as Wadi El-Natrum - see Saleh, et.al. 1972, p. 144) and quartzite, all crushed, are other sources of silica which may have been used, especially the first. Soda provides the flux and while not usually present (as in Egypt) in its natural state as natron is nevertheless readily obtainable from other sources. Calcium is found in limestone and chalk but it has been argued that lime was not necessarily a separate additive to ancient glass since the sands used in its manufacture could contain sufficient calcium oxide to account for the levels found (Sayre, 1965, p. 146). Shells are thought to be the source of this. It has also been suggested that glasses with a higher magnesium content may have been made with dolomitic rather than calcitic limestone (ibid.), but not enough analyses have been carried out to say whether this was an intentional selection of available limestones by the glass-makers of this period (Forbes, 1957, p. 115). It was only in the medieval period that the soda-lime glass gave way to the potash-lime glass and this explains the negligible amounts of potassium found and which represent impurities. Likewise the small amounts of aluminium are probably impurities derived from the sand used. Another probable impurity in the enamels (except brown) is manganese. Lead, tin, copper and iron where these are intentional, would be no great problem to obtain from scrap metal or the actual raw materials for metal-working if this was carried on locally. Cobalt ore would have to provide the blue colourant when needed since it is unlikely to occur accidentally in any of the other ingredients[90]. Antimony can be simply made by smelting stibnite with charcoal (Forbes, 1956, p. 47) and it seems likely that this was added as a separate component (Sayre, 1965, p. 147). Sayre (ibid. p. 148) also suggests that an oxidisation product of bronze may have been used as a colourant. It is unclear whether the well known balls of blue frit found on several Romano-British sites have any relevance as a colouring agent in enamel or even glass. Such a use has however been suggested for those found at Exeter (Fox, 1952, p. 8) and London (Marsden, 1980, p. 74). In any event it is clear that the raw materials for enamel would have been reasonably easy for the Roman-British enameller to procure. It should again be noted that glass could be used as cullet for the basic mix or specifically as a colourant (see especially Forbes, 1957, p. 179).

Little work has been carried out on the question of decay or weathering of enamel but in many respects the same internal and external destructive processes which cause the break-down of glass can also attack enamel. These processes are described elsewhere (e.g. Flower, 1880, pp. 65-165; Harden, 1956, pp. 333-334; Forbes, 1957, p. 112). In addition, of course, a too large or too shallow cell, the lack of any holding device (see next chapter) or poor application would not facilitate the retention of the enamel. Few decayed enamels have been scientifically checked. These all form part of the study on the British Museum's head stud brooches (Painter and Sax, _op.cit_). In four cases (_ibid_. nos. 1-4, pp. 155-159) the second totally weathered enamel was found to have been opacified with antimony which was said to suggest that the original colour was not yellow (since lead antimonate is a stable compound and would not be expected to break down) though it could have been white, blue or green. It is interesting to note in no. 18 (_ibid_. p. 167) that of the six cells still containing enamel two appeared to hold a green enamel and four a red enamel but analyses showed that in fact only cuprous oxide was present in all six and that therefore the only intentional colour was red. Likewise the colour of no. 19 (_ibid_. pp. 167-168) was not apparent but analysis showed it to be a weathered red. Both the enamels on no. 4 (q.v.) had been stained green with copper corrosion products but again analysis showed these had originally been red and another colour. This analysis is useful on weathered enamels and shows the caution necessary in dealing with a 'green' enamel.

However visual examination can be useful in giving some idea of the amount of decay or loss. Over the whole range of almost two thousand pieces of enamelwork examined there are approximately four hundred cases of decayed/weathered enamels and three hundred and fifty where the enamel is no longer in evidence. These figures refer to instances of one colour rather than to objects (which might also have surviving a perfectly recognisible colour). It may be stressed again that decayed or weathered enamels generally turn a green, white or creamy colour and care must be taken with enamels of these colours as a whole. Blue is found with a decayed/missing enamel twice as often as red is and since blue and red are by far the most frequent combination this may suggest red was prone to break down more rapidly than blue. Higher rates of decay were noticed especially on the group M brooches and the head stud (Q) brooches while low rates or better preservation were noted with dragonesque brooches, animal plate brooches and seal box lids. Some instances of poor workmanship may be pointed out. A number of enamels contain a few burst air bubbles and others have a rather bubbly appearance, both due perhaps to the paste not having been allowed to dry out properly before firing or too much heat having been applied. This can be seen on the seal box lid from Newstead (Curle, 1911, pp. 332-333, no. 5) and the button-and-loop fastener found at Traprain Law (Burley, 1955, p. 180, no. 234). On some the surface is covered with tiny cracks, the result of poor cooling, as with the white enamel on the oval and medallion plate brooch from Wickham Bushes (University Museum of Archaeology and Ethnology, Cambridge). The miscellaneous mount recovered at Traprain Law (National Museum of Antiquities of Scotland, reg. no. GVM344) provides a good example of one colour overflowing its cell edges into the next. There is of course the problem as to whether such pieces represent unfinished and discarded objects or not (and hence indicate local manufacture?). Meanwhile the metal base to which the vitreous substance is applied is examined next.

CHAPTER 7

THE METAL BASE

Both the type of metal chosen for the base and its subsequent treatment play an important part in enamel manufacture. This chapter examines the various metals suitable for enamelling and which of them were actually used during the period under discussion. Various analyses carried out on Romano-British bronzes are discussed. We look too at the actual cells created to take the enamel as well as the various devices employed to help keep it in place. The preparation of the metal and treatment such as tinning are included.

A great variety of metals are suitable for enamelwork and indeed have been so utilised (Shaw, 1972, pp. 31-39). The crucial and obvious point is that the metal must have a melting temperature higher than that of the enamel being applied to it. Since the actual enamel needs a temperature of more than 650°C to melt then those metals with a melting point of less than this are not suitable. Thus pure lead (327°C) or tin (232°C) could not be considered.[91] However the more desirable metals such as gold (1063°C), silver (960°C), iron (1535°C), copper (1083°C) and its various alloys such as bronze are suitable and have been used at one time or another during the history of enamel-working. Each is now examined in relation to Romano-British enamels though by far the most is said on the last, copper and its alloys.

Gold

One of the earliest known examples of enamelwork is to be found on a group of six gold rings discovered in a late Mycenean tomb on Cyprus and since then the qualities of this metal as a medium for enamel have been frequently used. Perhaps its greatest drawback is its relative scarcity and implications of high value. There is no example of its use from the Iron Age and only once does it occur among objects from the Romano-British period. Enamel is to be seen on the terminals of the gold bracelet from Rhayader now in the British Museum (Brailsford 1964, p. 14, no. 9, pl. III). Despite the frequent use of gold during the Anglo-Saxon period those pieces bearing enamel, e.g. the Dowgate Hill brooch (Jessup, 1950, pp. 125-126) or the Minster Lovell and Alfred jewels (Hinton, 1974, pp. 27-48), tend to be beyond the terminus of this present study. Certainly millefiori insets are found on the gold purse-lid, shoulder clasps and pyramids from Sutton Hoo (q.v.) though these are not strictly relevant.

Silver

This too is an easy and popular material to enamel and is usually more available and cheaper than gold. Its use is non-existent in the Romano-British enamel-working industry. Only the brooch found with the more famous Birdlip mirror may provide an early use of enamel on this metal. The 'eyes' at the top of the bow "seem at one time to have been concealed by a decorative inlay" (Green, 1949, p. 189) which only might have been enamel. For a greater use of silver it is necessary to look to Scotland to the post-Roman period when small numbers of such objects bear enamel decoration. The hand pins and plaques from Norries Law immediately spring to mind (Durham, 1864, p. 8) but there are other silver hand pins such as that from Gaulcross (q.v.) and the Pictish necklace terminals from Parkhill and Whitecleugh (q.v.). Also there is the mount on one of the bowls from the Saint Ninian's Isle treasure (Small *et.al*. 1973, pp. 54-55). It may be noted that in all these cases the enamel appears to be red only.

Iron

Equally suitable for enamel is iron. Iron seems to have been more commonly used as an enamel base on the Continent during the Iron Age (Dechelette, 1927, pp. 1053-1063, *passim*) than in Britain where it is rarely found. One may instance the Trevone brooch (Dudley and Jope, 1965, pp. 18-23), the bronze-coated iron Aldborough terret (Megaw, 1970, p. 157) or a small domed stud from Lochend Dunbar said to be iron and enamelled, though now in very poor condition (Longworth, 1965, pp. 180-181). There is too the Eastburn brooch (Stead, 1965, p. 115) and the iron frames on the Bugthorpe discs (q.v.). From Roman Britain have survived only the pair of iron plates, from dagger sheaths, found at Hod Hill (Richmond, 1968, pp. 137-138, no. 11 & Brailsford, 1962, p. 5, no. 5) but seemingly imported by the first wave of legionary invaders. Otherwise there appears to be no enamelled iron later than this from Britain. There is only the exceptional iron mount with red and yellow enamel found in an unknown context at Lagore Crannog in Ireland and hardly earlier than the sixth century A.D. (Hencken, 1951, p. 121).

Copper and Copper Alloys

From a purely visual examination, the over-whelming number of Romano-British and related enamels are applied on copper or one of its alloys. It may be permissible to note now that, since none of the analyses discussed below showed a pure copper to have been used, it is in fact copper alloys we are dealing with.

Most of the modern literature on enamel-working emphasises the importance of the base metal. Copper is the common one used nowadays on account of its being plentiful and cheap as well as being relatively easy to cut and shape - though its tendency to oxidise is a drawback (Theilade, 1972, p. 1). Copper was used in the experiments carried out for this report and was found to be satisfactory in all ways. Gilding-metal or tombac is highly recommended and has an advantage over copper in so far as it suffers from almost no oxidisation (*ibid*.). Also it gives a more attractive hue to transparent enamels though it does have the drawback that if fired more than twice the zinc burns out causing discolouration (Wenger, 1967, p. 7). Various figures have been noted for the amount of zinc required to make the alloy gilding-metal but it appears to range from 5 per cent up to about 20 per cent, the latter rarely being exceeded (Tylecote, 1962, p. 53)[92]. Higher amounts would seem to classify the metal as a brass which in one instance at least has been said not to be able normally to take enamel (Bates, 1951, p. 169)[93]. The latter however may only apply to the more industrial brasses (of the last two hundred years) with

30-40 per cent and indeed some brasses are said to contain as little as 10 per cent zinc (Tylecote, 1962, p. 53). A clear distinction does not always appear to be made between what is gilding-metal and brass.

Only recently has much been written on the make-up of Romano-British metals or to be more precise 'bronzes'. For a long time the only group of analyses available was that carried out by Smythe (1937, pp. 382-405) on twenty-seven small objects from the north, mostly from sites along Hadrian's Wall, especially Corbridge. The results showed both bronze and 'brass' (except in one odd case, the zinc averaged 12 per cent and therefore constituted gilding-metal rather than brass) to be in use and a more important distinction made was the use of lead for cast bronzes and its absence from the wrought bronzes and 'brasses'. Tylecote (1962, pp. 51-57) notes the almost complete lack of analytical material for Iron Age[94] and post-Roman metal finds and bases most of his discussion of the Romano-British metallurgy on Smythe's work. He suggests the widespread introduction of zinc into copper in this period; not only does it lower the melting point of copper from 1083°C to 1000°C but it gives the metal a more attractive yellow colour. Zinc itself is however difficult to produce[95]. Otherwise very little had been said on the subject and it might be added almost nothing on objects bearing enamel[96]. It was for this reason that a series of analyses was carried out on the metal of those objects, the enamels of which were being examined for the purpose of this study (Bateson and Hedges, 1975, pp. 187-188). Since then a very wide range of Roman 'bronzes' has been analysed and in a short note it was claimed to show that small bronzes are in fact brasses corresponding to modern gilding-metal (Craddock, 1974a, p. 254). This raised a certain amount of debate (Current Archaeology, 1974, pp. 349-350) where Craddock maintained that the overwhelming majority of small Roman bronzes are brasses. However the work is not yet published and again no enamelled objects were seemingly included. One point made and worth noting is the variation in the amounts of the constituents due in many cases to the re-use of scrap metal.

For this report the metal of thirty-five objects, all enamelled, was examined to check what it is made of and to see if the fact that the pieces bear enamel means they are different from non-enamelled metals. The objects have a wide variety of dates and provenances as already noted. The results for the amount of lead, tin, zinc and copper are to be found in the last four columns of table 3 of the paper cited. If we look at the amounts of zinc listed it will be seen that in many of the specimens it is, in fact, not present. Where it is, the amount varies between 8 and 20 per cent with an average of about 7.5 per cent and the great majority of objects have less than 10 per cent. We therefore appear to be dealing with a gilding-metal rather than a brass for the zinc-containing specimens. The large number of <u>bronzes</u> found to exist would make it seem that the introduction of zinc to the copper alloy was not so widespread as Tylecote (<u>op.cit.</u>) alleged and is at variance with Craddock's (<u>op.cit.</u>) contention that the majority of Romano-British small bronzes are brasses[97]. Lead is present in almost all the examples analysed and is on average 3 per cent but ranges from 0 to 7 per cent. This suggests, as does a visual examination, that all these objects were cast. It is unlikely that the one object (a trumpet head brooch of type S 11 - no. 5) with no lead falls into Smythe's wrought category. The presence of lead facilitates casting. On the whole the gilding-metal appears to have slightly more lead than do the bronzes. The amount of tin present varies from 0 to 15 per cent with an average of 7 per cent. However the bronzes have a higher average, 9.5, than the 5 per cent of the gilding-metal.

A major programme of metal analyses is now under way at the Ancient Monuments Laboratory in London on the brooches from Richborough. Over half of these, mainly of the first century A.D., prove to be brasses with up to 20 per cent zinc while the rest are bronzes with a high lead content (Musty and Butcher, 1976, p. 242). More recent detailed results suggest that in the first category the greater number of brooches were made with brass though alongside a 'significent' number in bronze but that afterwards bronze became the metal used for the manufacture of the majority of brooches. Over thirty of the brooches analysed belong to enamelled types and the composition of these shows much variation in the amount of zinc present (.5 - 20 per cent). The average moreover is double that for the non-enamelled types and in over a quarter of these enamelled brooches the zinc content exceeds 10 per cent (Bayley et.al. 1980, p. 246).

Generally then in the Roman period, two types of metal were used for enamelwork viz. bronze and gilding-metal, the latter usually referred to as brass. The latter has the advantage of not oxidising so much and possibly rendering the hue more attractive. Pure copper does not appear to have been used at all[98]. No groupings are as yet observable either in time or space with the objects analysed though the number so far may be too small to allow for this.

The Cells

We can now move on to the actual cells (sometimes called recesses or cavities) which receive the enamel. Enamelwork is generally distinguished on the grounds of how the cell is made, two types predominating viz. champleve and cloisonne. These are recent terms which have come to be applied to ancient enamels as well. Champleve enamel is where the cells are positioned beneath the surface of the metal and literally have been lifted out. Cloisonne on the other hand has the cells above the surface of the metal i.e. the cell walls are thin strips of metal soldered on to the surface of the object. This survey of the enamelwork has shown that all Romano-British enamelwork is of the champleve variety. While some techniques of the Iron Age do not fit either category,[99] the majority also are champleve as are those of the early Anglo-Saxon period despite the great amount of garnet cloisonne work. It therefore seems somewhat meaningless to say a Romano-British enamelled object is a champleve enamel though this is preferable to the erroneous use of the word cloisonne (e.g. Fox, 1894, p. 470; Megaw, 1970, pp. 173-174) or more particularly cloissons in referring to cell walls (e.g. Leeds, 1933, p. 24; Mottram, 1968, p. 259). Yet there is a wide variety of types within the enamels under discussion and these must be distinguished. It is proposed to do this on the basis of the actual enamel and its mode of being made to adhere to the base[100].

To return to the question of the cells which are all constructed out of the metal base of the object, are they literally 'champleve', cut or gouged out, or are they stamped or cast? Little work has been carried out on this aspect of the subject and indeed would require a greater examination of each object than time allowed. However a general visual examination gives the impression that the majority of the cells of Romano-British enamelled objects were made as part of the casting which as seen above appears to have been the method of manufacture for such small 'bronzes'. The addition of lead would have been of additional value here and the 'roundness' of many cells does not suggest the use of tools. Casting seems both the simplest and easiest method of making the cells in any case though this perhaps underestimates the skill of the bronze-worker with the gouge[101]. Jope (1955, p. 43) appears to accept that the Romano-British workshops used "the technique of preparing in the initial casting the pattern of recessed areas to receive enamel instead of tooling them out of the solid bronze". A scientific examination of the enamelled brooch from Wall showed that both the brooch and the cell for the enamel were cast (Werner and Organ, 1964, p. 46) and the same was found to be the case with the Southacre bracelet (Mottram, 1968, p. 259). On the other hand Faraday (Gage, 1836, p. 307) believed that the cells on the Bartlow vase had certainly been finished with a 'chisel' but was not sure whether this was merely secondary finishing or altogether the result of tooling. The former seems preferable though here certainly is one problem. Do tool marks or sharp edges indicate total use of a tool or only secondary work to complete the casting? Tool marks of some sort or another have been noted on several of the objects, in total between 5 and 10 per cent at a rough estimate. It may be worth noting a few of these. Several head stud brooches (group Q) show such signs as for example those from Caerleon (site museum) and Redhouse (Museum of Antiquities, Newcastle, reg. no. 1961, 19.10) where the bases of the cells of the bow have a few gouge marks. Similar signs can be seen on the dragonesque brooch from Charterhouse (Feachem, 1951, p. 40), the group RIV brooch from Caerwent (Museum and Art Gallery, Newport) and the S11 brooch from Barrington (University Museum of Archaeology and Ethnology, Cambridge, reg. no. 27.630). Many of the small simple cells on the group M brooches appear to have been gouged e.g. one from Mendip (Somerset County Museum, reg. no. 1083) and another from Woodeaton (Taylor, 1917, p. 108, Ashmolean Museum, reg. no. R41). Likewise with the non-brooch material there is evidence of tooling as on some of the seal box lids and miniature stools. Perhaps as good an example as any is the plaque from the Thames (Brailsford, 1964, p. 56) though here perhaps one should consider total tooling of the cells. However on the whole the enamel cells of the Romano-British objects seem to have been cast and in some instances finished off with some sort of graving tool.

For the Iron Age the usually accepted theory of the cells being gouged after casting (Leeds, 1933, p. 24 or Fox, 1958, p. 125) seems true from a visual examination of these more 'arty' objects of a less mass-production industry. To give but one example the cells - the edges of which show plenty of tool marks - of the Lakenheath terret (formerly the 'Suffolk' terret, q.v.) appear to have been cut after the object was made. It should be mentioned however that a small number do appear to have been wholly cast[102]. It might also be noted that in many of the spots decorating Iron Age objects (e.g. the terret from Colchester, reg. no. 596P.03 in the Colchester and Essex Museum) there is a small central depression in the base indicating the use of a flat-bottomed bit with a small central projection (May, 1971, p. 255)[103]. Drilling may therefore be added to casting and tooling as a method used to manufacture the actual cells intended for the enamel.

There is not much that can be said on the shapes and sizes of the cells. It is difficult to define the curving cells on many of the Iron Age objects and almost impossible to measure their area, in some cases quite considerable. Spots, with diameters of usually 3-5 mms. are frequent and later come triangles, squares and petals, often larger than the ones on Romano-British objects. In this latter period triangles are most common followed by lozenges and on a large number of objects a combination of the two. One may mention the large number of triangular cells, with sides 1-3 mms. long, on the domed plate brooches or the lozenges on many of the group M brooches. A combination of the lozenge and triangle is perhaps most frequent on the bows of the head stud (Q) brooches. Small squares and rectangles with sides varying between 2 and 4 mms. occur in some numbers on both the head stud and, to a lesser extent, on the M type brooches. Ovals or petals, curved triangles, crescents and tongues are found in smaller numbers. Small spots as well as spots and circles form the eyes of the dragonesque and animal plate brooches. There are too the circular cells on the flat round studs or mounts and some of the disc brooches; these vary in size according to the size of the object as well as the number of cells on it, for example on the circular stud from Chepstow (Brailsford, 1964, p. 56, no. 6) the large centre cell has a diameter of 2.1 cms. while the other narrow circular cells are 3 mms., 2.5 mms. and 3.5 mms. respectively in width. In addition one finds a great number of oddly shaped cells such as those following the form of particular types of animal plate brooches i.e. on the sandal shaped brooches where the cell is quite large or those formed by, for instance, the triskele-shaped septum on some disc brooches. It is often difficult to measure the depth of cells but they appear to vary from being a slight depression to being up to 2 mms. deep. Iron Age cells tend to be deeper than Romano-British ones which have a most common depth of about 1 mm.

Something must be said on what are termed here holding devices. This includes the over-used word 'keying' and deals with those methods employed by the enameller to help hold the enamel in place. This is a problem since some enamel applied to a flat piece of copper will only continue to adhere after firing if treated gently and in fact will often 'pop off' as the metal contracts on cooling.

A few points need to be made first however. If the holding device has been applied to the base of the cell - where 'keying' is generally assumed to be - we can only be aware of its presence when the enamel has been lost. Where in fact the device has failed to fulfil its function. It may be added that there are enough cases of missing enamel to make valid comments on these techniques. Secondly we have just seen that many of the cells are so small that they do not really need a holding device in addition to the cell walls which so act. Thirdly 'keying' in its normally accepted sense of scoring or cross-hatching on the base of the cell is not a method employed on Romano-British enamels despite a feeling current in the literature that it is. Even in the Iron Age it was not commonly used.

Among the best examples of 'keying' are the heads of the studs on the Thames (Waterloo Bridge) and 'North of England' helmets (Smith, 1925, p. 102). Here deep cross-cross incisions were made to help hold the enamel in place. However there are not many more instances of this that can be quoted and no example of its presence on the base of a cell. The nearest to this are the rows of tiny chisel marks on part of the cell base of the circular openwork mount with triskele design found at Dowalton Loch (Agnew, 1881, pp. 154-155) but this is an exception[104]. These are but two examples from the Romano-British period: both disc brooches with knobs on the edge and in both cases with incisions on the base of the outer cell. One is from Ixworth (Ashmolean Museum, reg. no. 1927.296) and the other from Corbridge (site museum). Mention might perhaps be made of those of the group M brooches which have an incised lozenge(s) on the upper bow though no trace of enamel survives on any and the probability is that these are purely decorative. In the post-Roman period the exterior print of the Loveden Hill hanging bowl (Fowler, 1968, p. 306) and the corresponding print on the bowl from Scunthorpe (Kendrick, 1941a, pp. 236-238) seem to be somewhat similarly treated. Otherwise one has to look to the Irish penannular brooches for the use of this device (Kilbridge-Jones, 1937, pp. 433-435, no. 60 but passim also). We have already mentioned the small proportion of objects (e.g. the Q brooch from Redhouse) with tool marks or rough bases (the Midlothian patera) but care must be taken here not to confuse the result of tooling, primary or secondary, with a deliberate intention to create keying. This is not always the right word for it even if it is deliberate.

A more obvious holding device has already been met with during the survey of Romano-British objects i.e. the toothed or serrated or notched edge to cell walls. This was found to occur on some of the miniature stools, the paterae, a few seal box lids and a small number of miscellaneous objects. It can be quite refined or rather crude and is not always on every possible edge. It is found on five of the miniature stools such as that from Chesters (Budge, 1907, p. 376, no. 655) and on most of the paterae, such as the handle from Kirkby Langthorpe (Butcher, 1977, p. 45, fig. 1), as well as two of the cups. The seal box lids from Camerton (City Museum, Bristol) and Winchester (City Museum) well display the technique. Among the miscellaneous objects so treated are the 'plate' from Silchester (Jope, 1955, p. 43) and a bronze plaque from Saint Albans (Frere, 1972, p. 138, no. 150).

Other such techniques, well displayed on the Midlothian patera (q.v.) are the use of vandykes, reserved laurel leaves and a reserved zig-zag line in a narrow channel. A fine example of the latter is seen on the Romano-British dolphin brooch with blue enamel from Lambay Island (Bateson, 1973, p. 68) though the same device seems purely decorative on the non-enamelled brooch from Saint Albans

(Richardson, 1944, p. 91, no. 1). The use of the zig-zag has been briefly discussed elsewhere (Jope, 1951, pp. 60-62). It may be added that some of the more decorative designs provide a useful, if not necessarily intentional, holding device i.e. the curves and scrolls of the Iron Age, the tendrils on some of the seal box lids or the triskele or wavy septa of the disc brooches and the spirals on the hanging bowl escutcheons. In some cases the devices seem to have worked well but in others, even with their presence, the enamel has totally gone.

Finally mention may be made of the silvering or tinning of a small part of the enamelwork corpus. It is generally accepted nowadays that it is tinning (e.g. Craddock, 1974b, pp. 349-350) rather than silvering - and tin with its very low melting temperature would be the obvious choice to add to the enamelled object. It would have to be applied after the enamel since otherwise the tin would burn off during the enamelling process. Two examples may suffice: the head stud brooch from Broxtowe (Castle Museum, Nottingham) and the animal plate brooch in the shape of a rabbit from Caerwent (Museum and Art Gallery, Newport) but there are many more mainly among the brooches and indeed the hanging bowl escutcheons. Where added this process would complete the work on the enamelled object.

CHAPTER 8

THE PROCESS OF ENAMELLING

Now comes the actual joining together of the two major parts of enamelwork examined in the previous two chapters. Here are examined various aspects of the process of making the vitreous substance adhere (by the use of heat) to the metal base thereby creating the true enamel. The technical viewpoint is explored first and then the general process of application, including the sources of heat and the tools used. This is followed by a study of the techniques employed by the Iron Age enamelworkers and then those employed by their Romano-British successors - the arrangement of the latter constitutes a suggested typology of the enamelwork of the period. Millefiori enamelwork is looked at in depth.

Let us first examine the technicalities involved in making an enamel[105]. It is desirable for the vitreous substance to form a permanent bond with the metal base. The former with its lower melting temperature must 'wet' the metal and form a bond with it (Shaw, 1972, pp. 156-157). It has been established that the Iron Age red enamel melts at about 686°C (Smith, 1925, p. 101) and while the melting temperatures of only Roman glasses have been scientifically examined (and found to lie between 700°C and 800°C) the enamels of the period with their larger amounts of fluxes should melt within the range 650 to 750°C - well below the melting temperatures of the metals used. Although the presence of lead could make the enamel more fusible it is more likely that the lead helps reduce the surface tension and viscosity of the enamel thus making it 'wet' the metal more effectively. At high temperatures chemical bonding takes place between the metal and the enamel and it is desirable to have the formation of cuprous oxide at the interface when the matrix is copper based (Shaw, 1972, p. 157). This point is note-worthy and is referred to again later in the chapter. The presence of copper probably helps in this but it is unlikely to have been added deliberately to the enamel for this purpose. In the last chapter it was seen that too much zinc in the base metal acts against a good bond taking place because of oxide development at low temperatures though some recent analyses (q.v.) indicate quite large amounts of zinc in the metal of some enamelled brooches. There should also be as close a coefficient of thermal expansion as possible for the enamel and metal so that the bond is not then broken when the object cools. The difference in expansion in fact causes the enamel to become compressed and since the majority of the cells on the objects under review are relatively small this may well help the enamel to stay in place. Loss of enamel is more common from the larger cells.

Before moving on to the actual process it may be as well to look at the tools or implements necessary to carry this out once the craftsman is provided with the enamel and the metal object. The most important tool to the modern enameller is the kiln. We do not know exactly what his Romano-British counterpart used in this respect but an oven would not have been a major problem bearing in mind the use of furnaces for pottery-making and more relevant perhaps, glass-making. An enamelling oven complete with iron door is said to have been found at Bibracte and is described elsewhere (Bulliot, 1899, p. 20) and if the suggestion that some of the furnaces at Wilderspool were for enamelling is correct then examples can be sought nearer home (May, 1904, p. 73 and plan III). Even if not, the latter were probably used for the manufacture of glass and possibly could have been used also for the production of enamelwork. An enamelling kiln is in any event quite simple to construct even without the benefit of bunsen burners. This assumes of course that a kiln of some sort was used but this was not necessarily so (Maryon, 1954, pp. 172-173). Experiments have shown (Tylecote, 1962, pp. 25-26) that an open wood fire can generate a temperature of 600°C-700°C but bellows or even a charcoal fire exposed to a natural March wind (giving an oxidising effect) could raise the temperature to 800°C though not consistently. This as noted above would be enough to fire enamels. The experiments just mentioned were in fact an attempt to smelt malachite; this was also placed on a flat pottery dish with an upright porous pot over it which thus reduced the oxidising effect of the wind; but in any event here was a primitive sort of kiln which would just be as effective for enamel. It has been suggested that in Ireland objects to be enamelled were placed on small circular pottery dishes (about 3 cms. diameter x 1 cm. high), which were placed on a fire and with the aid of a pair of bellows had a flame played on to the surface of the object to effect the adhesion[106]. This seems a perfectly feasible method. It is somewhat similar to one put forward for melting the metal for coin blanks in the hollows of the so-called 'coin-moulds' of the Iron Age (Clifford, 1961, pp. 146-147). Here the heat comes from a stock of charcoal flamed by bellows[107]. Again such a method could be used for enamel-working. Thus while a kiln is perfectly possible for enamel production in this period other methods may have been used.

Otherwise the tools needed are relatively simple. While modern enamels (see Figure 14A) can be procured in powdered form they also come in lumps and for these a mortar and pestle are essential. It is hardly likely that pre-powdered enamels were available in the period of interest. Hughes (1972, p. 105) has said that the red enamel of the Iron Age circulated in large lumps and sticks or rods of enamel have been found at the Villa D'Anthee (Henry, 1933, p. 68) and Lagore Crannog in Ireland (Hencken, 1951, p. 132). However mortars and rubbing stones are common finds from Romano-British sites and while it is not being suggested that this was a primary function, even for a small number, nevertheless they were readily available for the use of the enamel craftsman. Some sort of spatula is needed to help fix the enamel powder or paste in the cells but again a small knife blade or other such common implement is perfectly suitable for the job. A pair of tongs is necessary for placing and removing the enamelled object in and from the kiln but once more tongs (iron) are a commonplace find of the period (e.g. those from Traprain Law; Curle, 1915, p. 166, fig. 35,4) and even a pair of green withies is a possible substitute (Tylecote, 1962, p. 136). Finally polishing and grinding are again relatively simple and need no explanation. For the purposes of the enamels made for this study a carborundum stick and an 'abrasive polishing sponge' were used.

The one point stressed by all modern enamelling handbooks is cleanliness, both as regards the enamel powder and the object destined to receive it. Once the powder has been ground down from the sticks and lumps elaborate precautions are suggested to keep out dust and dirt. Likewise even finger-prints are to be strenuously avoided on the metal base. It is certainly agreed that these can cause flaws in the firing stages which could mar an object but I did not take such precautions during my experiments and do not feel that this hampered the end result. Likewise it is probable that the enamel-workers of Roman Britain did not go to any great lengths in this respect. In any event the enamel is applied either directly to the cells in the form of powder or mixed with a little water to form a paste and then so applied. The latter seems the preferable method especially in filling the angles of the cells more easily. It is of course necessary to let the paste dry out otherwise the water will boil off when placed in the heat. Powder or paste is applied with a spatula. If the enamel is in the form of solid slices (see below under complex and millediori enamels) then this is so much simpler and merely requires careful attention to positioning and care that the arrangement is not disturbed when placing the object in the oven. Nowadays a gum, which burns off without trace, is used even to hold powder in place before firing but this is really unnecessary and was hardly used in the period under review. Assuming a kiln is used and that it is at the correct temperature the object plus enamel is lifted by the tongs and placed in the oven for a few minutes. A spy-hole permits the enameller to see the enamel turning 'liquid' and once this is complete it can be removed and allowed to cool. The copper (used in the experiments) oxidises and becomes covered with fire-scale but this mostly flakes off and the metal edges and base can be polished and the enamel ground down if necessary. A short return to the heat restores the surface condition. If two or more colours are being fired they should, according to the text books, be fired in order starting with the one with the highest fusion point but in fact I found that various colours could be fired all at once without any ill effects - unless in the long-term. If the cell is deep a second application of powder/paste may be required. This may also be necessary to fill the corners. Sometimes a layered effect can be seen on ancient enamels and this may be due to the point just made. The firing process is therefore as simple as the tools needed to carry it out.

It is usually on the grounds of application that enamelwork is classified such as the modern ones on the grounds of their being champleve, cloisonne etc. Before attempting this on Romano-British enamels it is necessary to examine those of the Iron Age separately since many of these do not fit the suggested scheme. It has been said by Leeds that this was a period of experiment, hence the greater variety. Therefore the Iron Age techniques are treated separately.

There are two basic types of enamel in the Iron Age best described by terms borrowed from Jacobsthal (1944, p. 133) as <u>uplay</u> and <u>inlay</u>. In the former mode the enamel is applied above the level of the basal metal while in the latter it is found below the surface of the metal. There are various sub-divisions within each group. For the uplay group there are various methods of affixing the enamel but for the inlay group the variations occur in the size and shape of the cells and the colours used in these.

Uplay enamels of the Iron Age (Figure 10A)

1. a stud of red enamel held in place by a rivet through the centre - only a rivet hole in the metal, no form of setting (figure 10A, 1). The best example of this method is to be seen on the Bugthorpe discs. It is also used on the Eastburn fibula and was too the mode of holding the studs, presumably of enamel, on the Llyn Cerrig Bach plaque and the now lost shield bosses from the Polden Hill hoard.

2. a stud of red enamel held in place by a rivet through the centre and sitting in a circular depression or setting (figure 10A, 2). The stud on the handle of the Lexden cup and the missing studs on the Perth and Newstead button-and-loop fasteners belong to this variety. In many examples enamel and rivet are now gone and it should be borne in mind that where there is now an empty setting with a rivet hold this may have been intended for a rivet with an incised domed head (type 3) as on the tankard handle from the Seven Sisters hoard.

3. an applied rivet with large domed head which is incised to receive red enamel - set into a circular depression (figure 10A, 3). The studs on the North of England helmet and the tankard handle from the Seven Sisters hoard just mentioned have been applied in this way. The studs on the Waterloo helmet are slightly different in that the enamel has been applied to a disc, with incised head, which has been set into a depression and held in place by a rivet through the centre - the head of the rivet being above the enamel. Rivets only have been found at Hengistbury Head and Knob's Crook while the enamel shell only, once covering the rivet head, was discovered at Hyndford Crannog.

4. studs of enamel set in circular setting - no rivet and enamel above metal surface (figures 10A, 4). Probably best seen in the remaining studs on the Harpenden mounts and as far as can be ascertained on the Maiden Castle fibula. This method was probably also used on the Snailwell armlet, the Felmersham spout (eyes only) and the Colchester trefoil pendant. Where the enamel has gone and only the setting remains it is sometimes difficult to know whether the enamel was of this type or of type 7 where the enamel is flush with the metal surface and therefore inlay.

5. framework enamel - here the enamel is found inside a hollow openwork stud which acts as a framework to hold the enamel - the studs are in turn held in place on the object either by solder or a rivet through the centre (figure 10A, 5). Several examples of this type are known on the Battersea shield and the remains of the Hertford Heath and Lexden (?) caskets. The example from Hod Hill is of this type but shows a variation in that the enamel is used to hold two blue glass discs in place.

6. applied rivets with ring and dot of enamel sunk on the top (figure 10A, 6). Not strictly speaking an uplay but rather an inlay - however as it occurs above the surface of the object it is included here.

It is found on the Embleton scabbard where the enamel is preserved (red and ? blue) and possibly on the Bugthorpe scabbard too though there is now no enamel and perhaps there never was.

Inlay enamels of the Iron Age (Figure 10B)

7. simple spots of red enamel (figure 10B, 7). Red spots on their own are one of the simplest types of enamel and occur on a wide range of objects. They are to be found on the Birdlip mirror, the Cawston, Rickinghall, Snettisham and Stanton terrets and account for about 99 per cent of the enamelwork in the Stanwick hoard. Occasionally one finds other shapes in red alone, as the single small triangle on the button-and-loop fastener in the Middlebie board or the row of triangles on the Rattlesden terret. These may be included within this group.

8. relatively large cells with curving walls - generally in red (figure 10b, 8). Here the size of the cells is the main factor. The term 'cells with curving walls' is used here to cover the Celtic scrolls, lobes and trumpet patterns adorning much of the Iron Age metalwork. A good example of this type of enamelwork occurs on one of the quatrelobe harness mounts from Santon but it is also to be seen on the faces of the Lesser Garth terret, the Felmersham and Leg Piekarski bowl fittings, the Lydney Park stud and the Hod Hill sword hilts. Occasionally a second colour - blue - is found also in largish cells but alongside the red, as on the Colchester linch pin and Fayum and Eauze terrets.

9. relatively large cells with curving walls in conjunction with spots - former always red (figure 10B, 9):

(a) closely akin to type 8; red is the only colour used. A good example, with particularly large cells is the second quatrelobe harness mount from Santon. It is also to be seen on such objects on the Eckford cheek-piece, the Alfriston and Runneymede terrets and the trace hooks in the Seven Sisters hoard. It occurs in conjunction with small triangles - also red - on the quatrelobe harness mount from Polden Hill.

(b) here the spots are blue -whether glass or enamel has already been discussed - excellent examples are the Bapchild, Lakenheath and Richborough terrets.

(c) there are some examples with yellow spots. Three such examples may be given, the Auchendally terret and the London and Norton quatrelobe harness mounts.

10. small cells, which are angular, in red, blue and yellow (two or three of these) (figure 10B, 10). The cells are no longer large curving areas of red but have dwindled in size to much smaller triangles, squares, rectangles, lozenges, petals, ring and dot and various other shapes - angular for the most part - filled with a much greater variety of enamel than before. The group is quite large but should perhaps be subdivided. Group (a) would consist of enamels in all three colours, red, blue and yellow, mostly in a large number of square to rectangular cells as on the Embleton scabbard or the terrets from Saham Toney. However where the

enamel has decayed it is not clear whether three or only two colours have been applied, for example as seen on the two terrets and two bridle-bit fragments from Seven Sisters. On the other hand in some instances where all three colours are present - see the Silchester mount or cruciform mount now in Florence - the cells take on a different shape and better pattern than a row of simple squares. There is not therefore a great uniformity within this group. Group (b) includes objects with two colours in the combination of red and blue and red and yellow (these might form the basis for a sub-division as Leeds pointed out but his view that red with yellow is northern and red with blue, southern, does not quite hold good) found generally in a smaller number of cells though these have a great variety of shape. Good examples of this group are the red and blue enamel on the Rise bridle-bit and the red and yellow on the terret from Traprain Law. Red and yellow can be seen in much greater quantity as in the rows of inter-locking triangles on the belt plate in the Yorkshire Museum. The combination of yellow and blue is rare but does appear on the Thorpe sword hilt.

Leaving aside the Iron Age material we will now deal with the enamelwork of the Roman (and post-Roman) period and try to simplify it into some sort of order. Present day terms are usually applied to older enamels but as mentioned earlier such a system applied to Romano-British enamelwork would make it all champleve. It is reasonable to say that a Romano-British enamel is <u>per se</u> champleve. Yet there is a wide variety of types which need arranging. This can be easily and satisfactorily done on the way the enamel is applied and has to a large extent been hinted at already in the text. That is to say it is proposed to label an enamel as (a) simple, (b) complex or (c) millefiori.

<u>(a) Simple Enamels (Figure 11A)</u>

As the name suggests these are simple and straightforward to produce. There is only one colour of enamel to each cell on the object. This has been applied as a powder or paste. However there may be more than one colour, usually two or even three, on the same object but as long as these are confined to their respective cells the object stays within the group. We can then designate such enamels as S x 1,2 or 3 depending on the number of colours. Figure 11A shows what is meant and figure 14B displays some modern examples made in the experiments. Examples of this technique are to be seen on three finds from Newstead: a) a button-and-loop fastener - S x 1 = blue; b) a disc brooch - S x 2 = blue + red; c) a dragonesque brooch - S x 3 = red + blue + yellow (Curle, 1911, pp. 333, 330 and 320 respectively, pl. LXXXIX nos. 22, 13 and 8 respectively). The small and large domed plate brooches always have this variety, usually in the two colours, red and blue each confined to their small triangular cells. Dragonesque brooches, too, have exclusively simple enamels, being found with one, two or three colours from red, blue and yellow and so also do the trumpet head brooches (group R). The great majority of the head stud (group Q) and group M brooches as well as all the miniature stools and vessels along with most of the seal box lids belong to this category.

(b) Complex Enamels (Figure 11B)

Here two or more colours of enamel are found in the same cell without their being separated by a metal divide (septum). It is thus a more complex arrangement than that of the simple enamels. This type is often called 'mosaic' enamel but as the term has also been applied indiscriminately to millefiori enamels it is best perhaps to scrap the word altogether and have two new ones to describe these two different types.

There are various classes of complex enamels.

(i) In the first group pieces ('squarish') of two or three different colours are laid alternately side by side in a long narrow cell or in a narrow circular or oval cell. The great majority of the oval and medallion plate brooches display this, possessing one oval cell around the base of the raised central portion; in this oval cell (c. 8 mms. wide) are eight or twelve pieces alternately blue and white. The pieces themselves are approximately 9 sq. mms. Such is to be seen on the specimen from Lowbury Hill(Atkinson, 1916, p. 35, no. 34, pl. IX,34)and, in red and blue, on the example from Campsfied (Hunter and Kirk, 1952, pp.59-60, fig. 26,1). Along the bows of a pair of head stud brooches from Newstead are ten and nine pieces respectively, alternating between red and blue (Curle, 1911, p. 324, pl. LXXXVI, nos. 21 and 22). Much smaller pieces of this type of enamel are to be seen on the two or three long narrow cells on the upper bow of the small sub-group of M brooches (group M(e) - see above) such as the one found at Wilderspool (Thompson, 1965, p. 83, fig. 20, no. 4).

Again the idea is shown in figure 11Bi as well as on a modern reproduction in figure 15A, 1 & 2. What has been invariably noticed (see figure 15A, 1 and 2) is that the edges between two such colours are irregular but very well defined. This suggests that solid pieces or plaques (?cut from sticks or rods) rather than two pastes or powders had been used to create the pattern. The latter in any event would have been very difficult to apply in this fashion and would have resulted in a blurred edge. Compare the crisp edges between the colours in figure 15A, 1 with the blurred edge between the red and white (from pastes) in figure 15A, 3 and the point should be clear and acceptable. To make the former I used squarish plaques of coloured enamel purchased in this form but equivalent to pieces cut off rods (see figure 14, lower middle). Indeed the only case of such a blurred edge noted on the ancient enamelwork is that between the red and yellow on the escutcheons from Hitchin and 'near Oxford' (Leeds, 1933, pl. III).

(ii) The other major type of complex enamel is where a relatively large cell is filled with a background colour into which spots of another are inserted. This is well exemplified by the sandal ('animal') plate brooches with their frequent blue grounds and white or yellow spots e.g. those from London (Guildhall Museum, 1960, p. 4, no. 5, pl. 5, no. 5) and Wilderspool (Thompson, 1965, p. 83, fig. 20, no. 7). See figure 11Bii and figure 15A, 4. I suspect that the ground was first laid as a paste or powder and that the solid spots (pieces from small rods) of the second colour

were then placed in this and the two fired together. This seems to be the most simple method and this was how I made the modern reproductions - figure 15A, 4 - where the red spots are solid slices. However it is said that the spots in the blue enamel ground on the Wall brooch were inserted while the blue was 'still viscous but cooling' and that they were stuck to some support in the required formation for this purpose (Werner and Organ, 1964, p. 46). This seems a rather difficult (unnecessarily so) method and in any event surely the enamel would solidify too rapidly for it to be effective.
On the other hand the appearance of the white ground on a brooch from Woodeaton in the Ashmolean Museum (reg. no. R76) suggests that the spots had been pushed in while the white was still soft. Such a technique is also suggested for the five plate brooches with white grounds and black spots included in the group of brooches from Nornour recently examined at the Department of the Environment (Bick _et.al._ 1980, p. 57). There is thus a problem here as to how the spots were attached, but I still believe placing them 'cold' is the most sensible method and one which was probably most often used.

(iii) In addition to the last two basic types there are even more complex ones usually consisting of one or more circles of different colours around spots either as plaques as Bi or in grounds as Bii. See figure 11Biii. It might be argued that the former would be millefiori but it is not really so and does not appear as part of the Romano-British millefiori pattern repertoire. Finds with this technique are not common from Britain. Among these is the casket handle from the Walbrook streambed (Guildhall Museum, 1960, p. 8, no. 1) with two squares of a white ground with decayed spot and circle alternating with pieces of a blue ground with a white spot. A stud from Camelon (National Museum of Antiquities of Scotland, reg. no. FX339) has in one cell pieces of plain (?) yellow and pieces of blue each with a white circle and red spot. The centre of a brooch from Farley Heath (British Museum, reg. no. 53.4-19.56) has one cell with a blue ground in which are eight red spots, four with white rings.

(iv) Finally there is a small group of objects bearing a type of enamel which is best included in the complex group. This consists of a sulphur yellow ground speckled or sprinkled irregularly with tiny pieces (sometimes square, often irregular) of brown[108].
This technique is illustrated in figure 11Biv and by an actual example from Newstead (Curle, 1911, p. 329, pl. LXXXIX, no. 25). The latter provides an excellent specimen of the type having four large cells of such enamel around two central panels containing millefiori decoration. It is a solid rectangular belt-plate and is an unstratified find from near the west gate of the fort. Indeed this technique occurs almost exclusively on belt-plates such as the two from Caerleon (Site Museum) and those from Caerwent (Newport Museum) and South Shields (Bruce, 1885, pp. 261-262). Besides these pieces mentioned its appearance has been noted on a tear-shaped seal box from Leicester. Here there is no question of the brown not floating in the yellow. Probably it was easier to sprinkle the brown on the the unfired powder or paste of the ground rather than do this when the yellow alone was cooling. Curle (_ibid._) calls this a millefiori enamel but on one of those from Caerleon the brown has largely

disappeared leaving tiny, shallow, irregular holes with the yellow ground visible beneath - such does not constitute a millefiori inlay.

(c) Millefiori Enamels (Figure 11C)

Among some of the most outstanding small relics of Roman Britain are those bearing millefiori decoration. However this is a poorly understood process and indeed many do not even think of it as being part of the enamelwork repertoire. Yet it undoubtedly is, in so far as a vitreous substance - here thin slices of millefiori glass - is made to adhere to a metal base by the use of heat. The problem lies in the method of application.

There are two possible methods (figure 11c). Either the slices have been used as inlay in that they lie on a bed of red enamel which acts as an adhesive to hold them in place or they have been placed directly on the metal and fired in the same way as an enamel, being given enough heat to make them adhere but not enough to distort the pattern.

The arguments against the latter are that the millefiori glass would have had too high a melting temperature to be treated as an enamel and secondly the coefficients of expansion of the glass and metal differ too much to effect adhesion. Red enamel has a melting temperature of about 686°C (Smith, 1925, p. 101) and would act as an excellent fluxing agent. It was however seen above that the Romano-British enamels would probably melt between 700°C and 750°C and that the melting temperature of much Roman glass lies between these figures also. Thus the first difficulty is not really so great and the problem of expansion/contraction is unlikely to be relevant. It therefore seemed that the next stage was to try to make an actual straight millefiori enamel.

The experiments turned out remarkably well and quite respectable, straight (without red enamel under) millefiori enamels were made without difficulty. The complete process is shown in figure 15B. Number (1) shows six separate pieces of millefiori glass set directly on to the metal object. The next, (2), shows these beginning to join after a few minutes in the kiln and (3) is the position when they have coalesed and covered the whole cell. Here they are perhaps too distorted and this is a crucial point in the excerise, i.e. to remove them from the heat at just the moment when they have filled the cell fully but before the pattern is lost. An excellent example of this is to be seen on the roundal of the head-stall from Newstead (Curle, 1911, p. 297, pl. LXX1, no. 4) where the millefiori pattern in the outer circle is so distorted that it has cleverly been covered with a thin strip of embossed bronze which however has partly come away to reveal the craftsman's cover-up. To return to the experiment, number (4) shows the finished enamel half ground down and (5) completely ground. After a short return to the kiln the glossy surface is renewed and a complete straight millefiori enamel is shown in number (6). It might be added that it could be said that the pattern here looks as if there are flowerets floating in a blue enamel bed but such is obviously not the case. In any event it is easy to make such an enamel and it should be remembered that the experiment was hindered in having only circular pieces - there would be fewer gaps between square pieces and thus the cell would be filled that much quicker before distortion became a problem. Also the pieces were rather thick and needed a lot of grinding. Indeed I cannot see the need for a bed of red enamel to hold the millefiori in place. The only difference from ordinary enamel-working is that this technique takes two to three times as long (though still only a matter of minutes).

However even if this method is possible was it used on the many specimens of Roman enamelwork from Britain bearing millefiori decoration or are the pieces floating in a bed of red enamel? Henry (1933, pp. 67-69) argued that both methods were used though without providing any proof. Jope favoured the latter where the millefiori is held in place with a flux of red enamel (Wainwright, 1953, p. 68) and indeed this is the view that generally holds sway. It was to try to solve this problem that a group of twelve objects bearing millefiori decoration was chosen for special examination (Bateson and Hedges, 1975, p. 185). These are numbered as in the work cited - (a) to (j). Ideally what one would like to do is to obtain a section across an object with millefiori decoration, polish this up and examine it under the microscope: for obvious reasons this is not possible.

The ten objects examined are fairly representative of the patterns in use during the Roman period and the decoration remains in varying degrees of quality and quantity. It is important to note that in some cases red forms part of the intentional millefiori pattern - usually as a border (e.g. Appendix C - pattern no. A7) - and this must be distinguished from a bed of red enamel into which the millefiori pieces may have been laid, if such is the method used. In general a number of examples were observed for which there is clearly no red enamel matrix. For example clean metal could be seen through a very thin non-red enamel layer in (b), (h) and (i). Several cases were found where it appeared highly probable that red enamel was not used as a matrix ((c), (d) and (j)). Thin traces of red were found under the millefiori of (a), (d) and (e) and one plaque of (f). Even here it is conceivable that reaction with the bronze base could produce a thin interfacial layer of red CU_2O (see above). It appears however for (a) and (e) that the green pieces are set in red enamel. Thus it seems that the two techniques were indeed practised. Again it may be noted that the partial composition of the millefiori glass, in particular the lead content, (see below) is not so different from the more straightforward enamels. Unfortunately it is seldom that a certain answer can be given to the question of the use of a red enamel flux since a destructive examination is needed.

Since we can now include millefiori enamelled objects and know exactly what to look for, it might be appropriate to mention the number and types of such objects which have been found in Britain. Those noted (as far as the search went - see notes 4 and 5), and there are undoubtedly more, total one hundred and seven objects (including three from France) as well as the five instances of Anglo-Saxon hanging bowls which also bear this type of decoration.

The vast majority consist of flat circular studs, various mounts, disc brooches, generally of the type with a stud standing up in the centre, belt-plates and a few seal boxes.

A visual examination of these helps further with the application problem. There are certain rather clear cases of millefiori floating in a ground of red enamel. One might mention here the very important group Siii brooch from Kilpedher which now retains in its fantail only the bed of red enamel but which, when found, was covered by millefiori which then disintegrated (q.v.). Here then is an excellent example of millefiori enamel floating and where the red bed would not have been obvious. Somewhat clearer examples of this technique are to be seen on the escutcheons and plaques of the large hanging-bowl from Sutton Hoo (q.v.). A stud from Newstead (Curle, 1911, p. 332, no. 4, pl. LXXXIX, 17) has an interesting central cell of white (original or decayed red?) with just one piece of millefiori floating in it. Less clear examples of floating are to be seen on studs from Caerleon, Corbridge and Ditchley.

Rather than point out references for what are good straight millefiori enamels it might be better perhaps to give a detailed description of one such object viz. the stud from Chepstow (Brailsford , 1964, p. 56, no. 6, pl. XXl, no. 6). At first glance it would appear that in the centre are small plaques consisting of either nine or twenty-five white and blue chequers floating in a sea of red enamel. This is not so. In fact there are two types of plaques, (1) consisting of nine white and blue square rods with a red border and (2) twenty-five white and blue square rods with a blue border, laid side by side directly and so fused. Because the red is the stronger colour it predominates and catches the eye but the edges of the two types are quite clear; where some is missing there is no red layer underneath; there is very little red around the cell edge where one would expect to find it if it constituted a bed for the millefiori. The next cell is not a series of white flowerets floating in blue but a series of millefiori squares consisting of a pattern of a blue ground with white petals. Generally the blue has coalesed but in places the edge between two individual plaques is still visible. There is certainly no red at all in this cell. Although there is red in the next cell the same idea is evident i.e. squares of white and blue with an integral red border laid side by side and fired. Where the white or blue has gone there is no red beneath. Again in the outermost cell no red is at all visible and the pieces are one of a blue ground containing a white whirl. Such then is a straight enamel of millefiori.

As regards the millefiori itself, it is probably well known that it is rather like sea-side rock with the pattern running right through. It is made by arranging large coloured glass rods together in the desired pattern, fusing these together and then rolling or drawing them out (while still hot) into one long thin rod which however preserves the pattern intact. Slices (or plaques) can then be cut off the end. There is a wide variety of patterns applied to Roman objects and these are listed in Appendix C (see figures 12 and 13). These can however be reduced to four basic types:

A. consisting basically of nine squares

B. consisting basically of twenty-five squares

C. consisting of a floweret

D. miscellaneous such as the whirl or fern patterns

Patterns A7, B1 and C11 are by far the most common. The number of occasions a pattern appears is included in the list. Some are found only on the hanging-bowls. Blue and white followed closely by red are the most common colours; green, yellow and black are frequent, pink was noted also; there is much variation in the shading, especially of the blue. Again opaque rods are the norm but translucent ones are more frequent than in the straight enamels. Generally the pieces are square but circular ones were used, again in the Anglo-Saxon period.

The usual size of each piece varies between 2-4 mms. along the sides and .5-1 mm. thick. Often the cells to which they have been applied are quite large such as that at the centre of the Chepstow stud (q.v.) which is 2.1 cms. in diameter and contains no less than sixty-four pieces of millefiori glass. The three plain disc brooches (type 1) having only one cell are 2.5 cms. in diameter, containing seventy-eight squares (from Campsfield), 3.1 cms. (Corbridge) and 2.5 cms. again (Richborough). On one of the narrow circular cells of the mount from London there are nineteen pieces the edges of which are quite clear. On the brooch from Saint Albans the narrow circular cells are 3 mms. and 3.5 mms. respectively. The belt-plate from Corbridge contains four cells, two each 3 cms. and 1.8 cms. long by 3.5 mms. wide and 1 mm. deep.

Little work has been done on the actual rods making up the patterns. One analysis is of a piece of millefiori glass, which was in fact part of a vessel (Brill and Moll, 1963, p. 146 + table 3). Partial analysis of the millefiori enamels examined suggests that especially in the lead content, these are not all that different from the make up of the true enamels (Bateson and Hedges, 1975, p. 185).

We can say then there were two types of millefiori enamel applied during the Roman period viz. straight millefiori enamels (Ci) and floating millefiori enamels (Cii). This group as a whole has been rather neglected in the past and consequently often misunderstood. It is hoped that this has cleared up the subject to some extent.

(D) Simple enamels or glass floating in Red Enamel(?) (Figure 11D)

There is one further possible technique of application which must be considered. It has been suggested that some of the blue enamels are in fact merely a thin surface layer of enamel or glass floating on top of a bed of red enamel (figure 11Da). This may apply to some black enamels as well. The idea first seems to have been put forward by Francois Henry not as one of the main techniques listed (Henry, 1933, pp. 66-69) but slipped in somewhat later (*Ibid.* p. 90) as a follow-on from the floating blue glass spots of the Iron Age (q.v.). Jope believed this method to have been used on the blue petals of the Carlungie brooch (Wainwright, 1953, p. 68) and to have been widespread. It must be said as regards such a blue that no clear example was met with where red could be seen under the blue either where the latter has come away as one might occasionally expect or at the edges of the cell. There could be only two reasons for such a thing - technical necessity or scarcity of blue enamel. The former seems unlikely and the fact that blue is the most common surviving colour on Romano-British enamelwork rather diminishes the likelihood of the latter explanation. Yet a similar technique was used with millefiori enamel (technique Cii).

There is moreover the case of the black enamels. These are found on the type 11 plain flat disc brooch (with the one wavy six-knobbed septum) such as that from Wilderspool with black in the outer cell and red in the centre cell (Thompson, 1965, p. 83, fig. 20, no. 9). There are also examples known with a blue outer cell. Where this outer enamel has decayed or gone it could thus have been either black or blue. One of these was included in the group selected for analysis (Bateson and Hedges, 1975, p. 184, no. 11) and under the microscope a thin strip of red (50-100 microns) was visible around the edge of the black. Unless this is due to the desired formation of Cu_2O between the enamel and the metal (see above) then indeed it would appear that the black is floating. Because of the shape of the outer cell it is unlikely that the black is glass. It would have been difficult to cut glass to that shape but the black might be smaller pieces of glass which have coalesed though more likely it is a frit. The near absence of lead is to be noted here. If the black is floating then the fact that this variety of brooch also occurs with blue in the outer cell (e.g. Corbridge - Site Museum, reg. no. P. 232) may be significant to our problem. The study of the Nornour brooches now appears to add some evidence in favour of the argument for this technique (Bick et.al. 1980, pp. 60-61).

For the present answers remain unsatisfactory. It is perhaps wiser not to dismiss such a technique as unlikely though more evidence for its use is desirable. However it would be strange if the Romano-British enamellers had resorted to such a technique when, as we have seen, their other methods were perfectly sensible.

CHAPTER 9

MANUFACTURING SITES

It is perhaps unwise, bearing in mind the Villa D'Anthee, to attempt a chapter on enamel-working sites, yet it is a perennial question as to where the enamelled objects were actually made[109]. For the most part non-British sites spring to mind such as Bibracte (Mont Beuvray), the Villa D'Anthee or Lagore Crannog and it will perhaps be useful to have a brief look at these first. We will then discuss the various types of evidence available to suggest the presence of this particular craft on a site. This is used to suggest a list of sites in Britain where enamelwork was probably or possibly made and indeed to delete some alleged sites.

Perhaps the best known enamel-working site is on the hillfort of the Aedui on Mont Beuvray, more commonly known as Bibracte. It is situated about twenty-five kilometres from Autun (Saone-et-Loire) and was extensively excavated during the latter half of the nineteenth century by Monsieur Bulliot. An impression of the finds may be gained from Bulliot's own description of the enamelling area, "Tools lay all around; the furnaces were still full of charcoal; besides finished articles, were others hardly started and yet others in the middle of manufacture, one was still covered with baked clay; all around were fragments of raw enamel, clay crucibles, sandstone polishers, a considerable quantity of wasters, blisters, refuse from the trimmings, vitreous shells which retained the imprint of the bronze designs..." (Bulliot, 1899, p. 5). There is thus the complete range of evidence - enamel, objects and tools - available from Bibracte (ibid. pp. 6-42). Perhaps the most interesting is the actual red enamel in its various states as a raw material, discarded as waste or flaws and as the finished product. It may be noted that Bulliot believed that the enamel was manufactured on the site (ibid. pp. 36-37) and the fragments of quartz found in hut 18 (ibid. p. 20) could have been the source of the silica; no large blocks or lumps of red enamel, as might have been imported, were apparently discovered on the site. Also of note is the fact that the objects to be enamelled were made at Bibracte as well and were found in the various stages of manufacture as bronzes as well as objects in the process of being enamelled (ibid. pp. 18 and 127). So the three stages of enamelwork - the metal, vitreous substance and the application of the two - are clearly seen here. The tools of the trade are well represented. The major item is of course the furnace and here one can be reasonably sure that its function was for enamel-working. One in hut 18 was circular, one metre in diameter, sunk half a metre into the ground and was filled with a mass of baked clay presumably from the collapsed dome; along with this was found the iron door, a thin square of 15 cms. with evidence of hinges (ibid. p. 20); other furnaces were unearthed in huts 20 and 23. In the latter were also found two broken clay tuyeres, an iron poker and a small thin iron shovel (perhaps for lifting objects in and out of the oven) along with the usual metal and enamel debris. Tongs also were recovered as well as an iron tripod, chisels, files and hammers. In addition to a small quern-stone were many crucibles - some with a red vitreous skin on the exterior. The most curious tools were pieces of a fine-grained sandstone with hollows corresponding to the various sizes of the enamelled studs and presumed to be used in the polishing process. Bulliot (1899, p. 29) thought the grain size too rough for complete polishing and suggested that the small perforated wheels of very fine baked clay somehow finished off this process.

For a long time the star enamel-working site of the Roman world was the Villa D'Anthee. The Villa D'Anthee is situated in the province of Namur in Belgium and again was excavated in the late nineteenth century by Monsieur Grosjean and was published after his death in the Annales de la Societe Archaeologique de Namur by Monsieur Marmol between 1877 and 1881. However it was not until the turn of the century that the description of an enamel workshop was forcibly put forward (Bequet, 1902, pp. 248-256). The idea was quickly accepted (e.g. May, 1904, p. 77) and Henry (1933, pp. 123-128) not only repeated the 'evidence' but assigned a "production d'une extreme variete" to the workshop. It is only within the last few years that doubt has been cast on the whole idea of enamel production at the Villa D'Anthee (Spitaels, 1970, pp. 209-241). This is based on a complete lack of such an idea in the manuscript notes of Grosjean or the publications of Marmol as well as a hypothesis of Bequet of 1893 becoming fact by 1902. In addition there is the grave lack of archaeological evidence among the finds now housed in the museum at Namur especially the fragments of crucibles with traces of a substance still adhering (Bequet, 1902, pp. 252-253) but which analysis proved not to be enamel (Spitaels, 1970, pp. 224-225). The pendulum of opinion has now swung in the opposite direction and the site is no longer included as a centre of enamel production. This is to the good in so far as the problem of origin can no longer be solved by a too fascile assigning to the Villa D'Anthee yet it seems to ignore the evidence for enamel-working in some degree even if the site can no longer be regarded as the main supplier to the Empire. There are at least the four rods of glass (Henry, 1933, p. 68, pl. 1) as well as the greenish vitreous powder and brown enamel which were analysed and discussed as raw enamel (Spitaels, 1970, pp. 240-241); again one should note the blue frit. Grented that there are few enamelled objects from the site (presumably finished?) and that the tweezers, compass and punches (or styli?) like the rest of Bequet's evidence are questionable nevertheless the Villa D'Anthee should not perhaps be dismissed entirely from a list of enamelling sites. The main lesson to be learned is that evidence for manufacture needs to be carefully examined for each possible site.

We might next look briefly at two post-Roman sites in Ireland of interest to the question of manufacture. The first is Lagore Crannog in County Meath excavated 1934-6 (Hencken, 1951, pp. 1-247). Here there is evidence for bronze-working, glass and enamel production. The evidence for bronze-working includes pieces of copper ore, moulds, crucibles, tuyeres, scrap and waste metal and bone trial pieces (*ibid.*

pp. 8, 126-127, 170, 234-240). Moulds for glass studs, one actually with the stud still in it, and fragments of glass, suggested to be imports for use as cullet, were also discovered (*ibid*. pp. 2, 127-129). However perhaps the most important discovery was the rod of millefiori glass and the two rods of opaque blue glass which it was hinted might have been intended for making up another rod of millefiori glass (*ibid*. p. 132). Various enamelled objects, some with millefiori, were found on the site, the belt-plate of iron previously noted being of special interest. The second site is a large rath at Garranes in County Cork. Here the excavations again produced evidence for a metal, glass and enamel workshop (O'Riordain, 1942, pp.77-150). Moulds, crucibles, unfinished penannular brooches and pins, a casting jet and ingots point to bronze-working (*ibid*. pp. 93-100, 107-110, 121-139). Mortars and tongs were also found. More important perhaps are the fragments of millefiori glass and enamel. Three pieces of millefiori glass were found, one being only a few millimetres long but said to be possibly the very end of a rod, a somewhat larger piece of a square rod with a simple pattern in red, blue and yellow (the red forming a border) and most exciting of all a circular piece of rod still fitted into the end of a bronze tube obviously meant to hold it while slices were being cut off (*ibid*. pp. 118-120). Again various fragments of glass were recovered (? cullet) as well as three glass rods and several small pieces of red and green enamel - some of these found fused to a clay crucible (*ibid*. pp. 120-121).

Before proceeding mention must be made of the Romano-British site at Nornour which for a time appeared to be about to become a British Villa D'Anthee[110]. The site, on the Scilly Isles, was excavated between 1962 and 1966 when the most spectacular find was the large group of brooches, most of which are enamelled. The hypothesis was advanced that the majority were actually manufactured on the island (Dudley, 1967, pp. 28-64). However investigation both in the field and the laboratory by Miss Butcher and her colleagues has led to the conclusion that the site is not a factory but rather a shrine at which the brooches were deposited as votive offerings (Butcher, 1977, p. 44 and Biek *et.al*. 1980, p. 51). Useful as it would be to have the location of a major enamel-working 'factory' in Britain, Nornour is not such a site.

Since we will not be fortunate enough to examine a site in Britain with so much evidence as Bibracte it may be appropriate to examine the various kinds of evidence to look for in trying to establish the presence of enamel-working on any particular site. Raw or scrap enamel is perhaps the best indicator of enamel production on a site and one should include rods of millefiori glass and opaque glass/enamel as well perhaps. It may not always be possible to recognise cullet, i.e. scrap glass, as opposed to discarded broken fragments (refuse) and in any event the cullet could be for glass-making as much as enamelling. The same applies to raw materials such as flint or quartz. Bronze-working on a site does not of course imply enamel-working but if the objects being made are types normally bearing enamel then the examination of the foreign sites would tend to suggest that the application of the enamel took place in the same workshop; the types of object being made may be gleaned from the remains of moulds, lead models and unfinished or flawed castings. Unfinished or poorly made (and therefore discarded) enamelwork would also point to manufacture at its find-spot. Tools - i.e. furnaces, crucibles, mortars, 'spatulae', tongs, chisels or polishing stones - on their own cannot really be pointers since such are common to bronze- and glass-working but in conjunction with for instance an unfinished object would add complementary evidence. Since red or green vitreous coatings have sometimes been said to indicate enamelling it might be as well to say these can result from metal-working or even just contact of a crucible with heat (Tylecote, 1962, p. 137; Alcock, 1963, p. 140)[111]. So on its own this cannot be a reliable guide though such red coatings were noted on some of the Bibracte crucible fragments. Mention should also be made of 'blue frit'. It is not always clear whether this is a blue vitreous substance given the general term 'frit' instead of enamel or glass or whether it is the balls of a blue grangular chalky substance often found on Romano-British sites. The latter is sometimes stated to be colouring matter for enamel but is more generally considered to be a paint pigment (see Atkins, 1971, pp. 56-58 for references and discussion). Finally there is the question of large numbers of enamelled objects from the one site suggesting manufacture there but as with Nornour extreme caution must be exercised with this argument. One must be careful not to suggest manufacture on a single indicator only unless a careful examination warrants such an assumption.

There is not such a lack of evidence for enamel-working in Britain as might be expected. Certainly there is no site of the calibre of Bibracte nor indeed many sites where the evidence is particularly strong. Yes it is possible to point to some where one can be reasonably sure that the craft was carried on in some degree. There are others where it has been claimed that enamelling was present but where the evidence does not really bear out such a contention. Again there exist certain sites where it is possible such activity took place. We will look at each of these categories in turn listing such sites which, while mainly of the Roman period, do cover the complete span of years under examination[112].

Probable manufacturing sites

1. Colchester

The excavators thought it probable that enamelling was carried on at Colchester on the basis of finds of pieces of blue and red frit (Hawkes and Hull, 1947, pp. 345-346). A crucible from Butt Street is said to bear 'distinct traces of red enamel adhering to the sides' (Bulleid and Gray, 1911, p. 303) but this along with other crucible fragments may merely indicate bronze-working.

2. London

London too is often quoted as a centre of enamel-working (Merriefield, 1965, p. 87) and here an unfinished object provides the best evidence for this. The piece is the unfinished plaque recovered from the Thames and already discussed in chapter 4 (Brailsford, 1964, p. 56). 'Crude lumps of green enamel' are said to have been found in Nicholas Lane (Wheeler, 1946, p. 32). Finds of balls of blue frit "used in enamelling" have been made in Cheapside, Lombard Street and Aldermanbury (Marsden, 1980, p. 74). Although these may not all be valid evidence there are some pointers to the process having been carried on in the capital.

3. Traprain Law

Perhaps the best piece of evidence for enamelling on this site is the square bronze ornament divided into four triangles, two with yellow and two with blue enamel. One of the cells containing yellow enamel has not been completely filled; the enamel has overflowed from the cells; the enamel appears bubbly and unpolished. The piece has rightly been regarded as a failure by the excavators (Curle and Cree, 1921, pp. 187-188, fig. 21, 13 - good illustration) and others (Burley, 1955, p. 194) who considered it as an indication of enamelling on the hill. Burley (ibid. p. 135) thought the thin-walled triangular crucibles, some bearing slight traces of molten glass in the bottoms, part of the enamellers' equipment but as she notes in relation to the iron tongs these could have been used for glasswork (beads and bangles) as well as enamelwork. Perhaps as useful are the moulds particularly those for button-and-loop fasteners (ibid. pp. 219-221, especially nos. GVM 556 & 7) though none of those found is for an object bearing enamel unless the fasteners were meant to have the cells gouged out subsequent to casting. Such objects as the numerous small tools of iron and bronze and the stone rubbers and polishers could have been employed by the enamellers as well.

4. Wilderspool

May (1904) went so far as to include 'Enamellers' Workshop' in the subtitle of his work Warrington's Roman Remains. The so-called 'jewellers' workshops' provided evidence for bronze- and enamel-working though the former seems more certain than the latter. The finds include 'coloured enamel paste or frit', 'three amorphous lumps of coloured enamel or frit' and 'a small angular fragment of black, glassy, unaltered enamel or frit' (May, 1904, pp. 73-75). Equally interesting is the lead model for a trumpet brooch (group Rii) though the bow and head are plain (ibid. p. 75). Nevertheless it suggests the manufacture of a type of object which often bears enamel. On the other hand there do not appear to be any moulds or flawed castings from the site though there is an unfinished bronze ring (ibid. p. 74). Otherwise while there are the furnaces and tuyeres, crucibles (some with vitreous coatings) raw materials such as sand, flint and chalk, tools, cullet, slag and waste products, these could be connected solely with bronze and more especially glass manufacture which were undoubtedly carried on to a great extent. Thompson (1965, pp. 67-87) in his re-assessment of the site starts by declaring "there is evidence for....the enamelling of metal objects" (ibid. p. 72) but "this cannot be regarded as certain" (ibid. p. 76) though he then goes on to suggest that the presence of enamelled brooches indicates this activity (ibid. pp. 83 & 86). Although caution is necessary it does seem likely that enamelled objects were made at Wilderspool.

5. Brough of Birsay

The most interesting find from here, among the scrap glass (?cullet), crucibles and moulds, is the lead model for a hanging bowl escutcheon of the type with the trumpet spiral decoration, usually bearing enamel (Curle, 1974, pp. 301-307).

6. Craig Padrig

One half of a mould for a hanging bowl escutcheon, with enamelled parallels, has been found here recently (Stevenson, 1972, pp. 49-51).

7. Dinas Powys

Scrap glass, scrap and waste bronze, part of the lead model for a penannular brooch, rubbing stones, polishers and various rods of glass including millefiori point to bronze-, glass- and enamel-working on this post-Roman site of the fifth to sixth centuries onwards (Alcock, 1963, pp. 53, 104, 120-122, 140, 160, 162, 187). The excavator was later somewhat more certain of the presence of enamelling here and noted the scrap glass for re-use as beads, bangles and "as a major constituent of the enamel inlays for fine brooches" (Alcock, 1971, pp. 233-235).

Possible manufacturing sites

8. Brough-under-Stainmore

A large number of small bronzes, including many flawed and unfinished specimens, are said to have been found here and it is usually accepted that they were cast *in situ* (Birley, 1958, p. 42). "Flawed castings of trumpet brooches are fairly common at Brough-under-Stainmore but though Q brooches were made there I have never seen a flawed casting of one" (Collingwood, 1930, p. 56). The bow of a buckle and a seal box lid, both enamelled, are said to be unfinished (Collingwood, 1931c, pp. 83-84). More recently this evidence has been regarded as 'highly deficient' and a lead model for an unspecified type of brooch given as the best indicator for manufacture (Boon and Savory, 1975, p. 43).

9. Caerleon

Rods of opaque blue glass as well as a ball of blue frit have been recovered here and are in the site museum.

10. Caerwent

Excavations in 1908 in house XIX N produced pieces of light blue frit "for use on enamelled metalwork" and similar frit found previously "seems to indicate that the beautifully enamelled objects found on the site were probably made there" (Ashby *et. al.* 1910, p. 12). This frit could refer to balls of blue frit not necessarily intended for enamel.

11. Canterbury

Excavations in Saint Margaret's Street have "produced evidence of a possible enameller's or bronze-smith's workshop"; this includes crucibles containing 'enamel' (Goodburn, 1979, p. 336).

12. Charterhouse

Charterhouse is included in this list on account of the large number of enamelled group M brooches (especially the variety with the lozenge-shaped expansion in the middle of the bow) found here. While Nornour provides an argument against this form of evidence nevertheless the great variety of types from there contrasts strongly with the basically single type from Charterhouse, probably manufactured and enamelled *in situ*. Crucibles have been found here (Bulleid and Gray, 1911, p. 303).

13. Chichester

Examination of vitreous deposits on various crucibles found in Chapel Street suggest they were for the manufacture or use of sealing-wax red glass or enamel; a strong signal was given for lead and cuprous oxide appears to be the colourant (Bayley, 1978, p. 254; and also see Butcher, 1977, p.68, note 3)

14. Kirkby Thore

Collingwood says that traces of the manufacture of trumpet brooches were found here but he notes the same for Traprain Law (Collingwood and Richmond, 1969, p. 297). Again the evidence has been queried recently (Boon and Savory, 1975, p. 43).

15. Mote of Mark

Much evidence for the manufacture of bronzes, including penannualar brooches, has been found here in the form of moulds and crucibles (Curle, 1914, pp. 140-147). Fragments of glass are generally regarded as being cullet for use in the production of enamel or other vitreous inlays (Laing, 1975, pp. 101-102; Longley, 1980, p. 4).

16. Shakenoak

A small amount of a vitreous substance found here has been described as enamel bearing a close resemblence to the lump of sealing-wax red glass enamel from Tara (Brodribb *et. al.* 1978, p. 94). If this is so it may indicate enamelling on the site.

17. South Shields

What appears to be an unfinished casting of a trumpet brooch of group Rii type may point to the manufacture of enamelled types of brooches at South Shields (Museum of Antiquities, University of Newcastle, reg. no. 1925.35 for a group).

18. Stanwix

Chance finds of over a dozen flawed castings, unfinished castings and scraps of waste metal have come from here and other evidence of a metal-worker's shop includes a square-headed button-and-loop fastener - "an unfinished casting never touched by a tool - intended for enamel but not begun" (Collingwood, 1931a, p. 41 and 1931b, p. 73). Doubt has however been cast on the workshop theory (Birley, 1931, p. 62).

19. Woodeaton

"It is clear that the site was a centre for the production of fibulae and other bronzes" (Taylor, 1917, p. 99). If this is so then perhaps some of the enamelled objects from Woodeaton were actually made there.

However the evidence "may point to a temple or shrine but possibly manufacture or a fair" (Kirk, 1949, p. 6). Certainly there is some evidence for bronze-working (*ibid*. 28).

20. Wroxeter

One of the rooms of the baths insula was formerly interpreted as an enameller's workshop though there is no hard evidence for this (Webster, 1975, p. 6). A ball of blue frit has also been found here (Bushe-Fox, 1913, p. 10). Among the extensive evidence for glass-making in Wroxeter (e.g. Wilson, 1973, p. 287) there may lurk indicators for enamelling and the site should not perhaps be too readily dismissed.

Doubtful manufacturing sites

21. Exeter

It has been said that "the native craft of enamelling was carried on" in Exeter (Fox, 1952, p. 16) but the evidence rests only on the balls of blue frit "which had been used probably for enamelling" (*ibid*. p. 6) and fragments of crucibles with a vitreous coating containing lead and more likely due to the melting of that metal than enamel (*ibid*. p. 64).

22. Heronbridge

The suggestion has been made that some of the furnaces were "perhaps for preparing enamels" (Hartley, 1954, p. 3) but the vitreous glaze on the crucibles appears to be due to bronze and the moulds are for objects which do not normally bear enamel. (*ibid*. pp. 5-6).

23. Meare and Glastonbury

The finds from these Iron Age sites are said to include "the workshop debris of bronze-smiths and enamellers" (Frere, 1974, p. 69) but a study of the excavation reports does not support the latter. "The absence of objects ornamented with enamel" has been noted at Glastonbury (Bulleid and Gray, 1911, p. 31) and the red glaze on the crucibles from Meare seems to have resulted from metallurgical use (Gray and Bulleid, 1953, p. 255). Glass beads do however seem to have been manufactured at Meare (Guido, 1978, p. 16; Henderson, 1980, pp. 60-61).

24. Nornour

It has already been said that Nornour is unlikely to have ever been a manufacturing site (see *supra*; Butcher, 1977, p. 44).

25. Richborough

It was said that a bronze brooch "straight from the mould and unfinished" might suggest manufacture of such objects (Bushe-Fox, 1926, p. 43, no. 5) though corrosion prevented its being declared enamelled or not. Conservation work has now shown it to be indeed enamelled but also to be quite finished (Musty and Butcher, 1976, p. 242). Other evidence for bronze-working and balls of blue frit were also found.

26. Saltersford

Preston (1916, pp. 48-49) suggested enamel-working here on the rather meagre grounds of waste bronze and a chisel.

27. Victoria Cave

Likewise King (1971, p. 414) seems to suggest that objects found here were enamelling tools but it seems unlikely that the craft was carried on here.

There are therefore many sites which can be at least discussed as manufacturing sites for enamelwork. None is outstanding but several provide reasonable evidence for the craft having been carried on in Roman and post-Roman times. Others have been put forward as such sites but like Nornour the evidence does not stand up to close scrutiny. Several more sites provide pointers towards the possibility of such activity. Unfortunately the evidence required is of a nature that can only suggest enamelling as one of three possible crafts (bronze and glass being the other two products) and unless raw enamel or better still an unfinished enamelled object has been found - and recognised - we must be satisfied with much less than Bibractian conditions. On most of the sites mentioned enamelled objects have been found, sometimes in large numbers, and these in conjunction with other evidence lend weight to the hypothesis of manufacture. Unless there is some idea of what was being made bronze-working sites as such have not been included and likewise the presence of glass-working, while it would facilitate enamelling, does not in any way imply the latter.

CONCLUSIONS

The major aspects of the Romano-British enamel-working industry have now been looked at to a greater or lesser degree. The products have been examined in some depth as well as those objects which were manufactured in the periods before and after, and indeed form part of the one story. Then came the techniques of actually making the objects: the vitreous substance or enamel itself, the base to which it was applied and finally the methods and ways of making the two adhere. An attempt was made to list those sites where such activity may have taken place. These and other points soon to be covered help build up a composite picture of the activities of the Romano-British enameller.

We are now in a better position to define just what products came out of the ancient British enamelling workshops. The Iron Age provides a varied but not numerous range of material associated mostly, as has often been pointed out, with the weaponry and horse or chariot fittings of chieftains and warriors. Sword and scabbard mounts, helmets, shields and belt-plates have been found along with terrets, bridle-bits, check-pieces and assorted mounts in much greater numbers than brooches, mirrors or studs which bear enamel. The corpus is greatly enhanced by the enamelwork included in the six major hoards already discussed. Yet if less elaborate, the enamels of the Romano-British period show a great increase in numbers and a quite different range of material bearing the decoration. Brooches become the chief medium for the craftsmen who applied their art equally to bow and plate brooch. Among the former the head stud (group Q), the trumpet (group RII and IV and IVa), the group M and the group S types provide the majority of enamelled specimens. A few dolphin, knee and thistle brooches are similarly treated. The dragonesque group is probably among the best known of the plate brooches bearing enamel but what has been termed the domed plate brooch is a frequent find. Likewise the oval and medallion brooch provides a small class of Romano-British enamelwork. The term disc brooch has been restricted to plain flat circular plate brooches of which the enamelled examples have been divided into sub-types based on the design of the enamel cells. Many more of these were probably made in Britain than has hitherto been accepted though indeed the group has been poorly examined in the past. Likewise some order has been given to the animal plate brooches and again there seems little reason to look across the Channel so readily for their source. There is no reason why the cocks, hares, dogs, sandals and horse and rider brooches should not have been manufactured and enamelled in Britain rather than Gaul.

Leaving aside the brooches, the other material proves to be numerous, as well as covering an interesting range of small objects. Button-and-loop fasteners are perhaps the best known type to have emanated from the workshops in Britain although only a few sub-groups actually bear enamel. In addition the enamelled paterae and cups include that interesting group of Wall souvenirs. Noteworthy too are the miniature stools and belt-buckle plates of long narrow rectangular form. Studs and mounts come in a multitude of shapes and sizes, many of which may have been made in Britain. The same may be the case for one of the largest groups of non-brooch enamelwork viz. the seal box lids. It is unlikely that all these represent ancient imports. Likewise some finger-rings and such exotic pieces as the Thames plaque may be regarded as Romano-British. There is a dramatic drop in the amount of enamelwork which has survived from the post-Roman period. This includes a very small number of brooches, pins and miscellaneous objects still displaying enamel decoration but the bulk is found on the hanging bowl mounts, especially the escutcheons. Thereafter the ancient enamels found in Britain are few and of a different technique thus spelling an end to an industry which produced a wealth of products for over seven hundred years.

From a visual examination of this material, in conjunction with the published analyses, we are now in a much better position of knowing how the enamellers went about their business. We can break loose from blithely talking about 'champleve enamel' and see that such a phrase covers three quite specific aspects: the actual enamel, the base to which it is applied and the method of application. In discussing the last of these a new system for describing enamelwork more fully and accurately has been put forward.

The opaque red enamel of the Iron Age is now well known following the recent work of Hughes. We still have to rely on the nineteenth-century analyses for a complete break-down (the basic glass is of the alkali-lead-silica type) but Hughes has shown that the presence of lead acting on the copper in reducing conditions results in feathery red crystals of cuprous oxide which give the enamel its brilliant sealing-wax red colour. This red is by far the most common colour on Iron Age objects, generally found alone until gradually it is accompanied by blue and/or yellow in spots or small geometric cells. The blue in some or all of the spots is a glass, lacking, in the case of the Lakenheath terret at least, lead and owing its colour to cobalt or copper but where the blue is a true enamel we are as yet ignorant of its make-up. The same is unfortunately true of the yellow.

The amount of information available on Romano-British enamels has been recently greatly extended. Not only has a much greater amount of enamel been subjected to analysis but it has been looked at in more detail than previously. The basic enamel was found to be similar to normal Roman soda-lime glass but has, like the opaque red Iron Age enamel, the interesting addition of lead in all cases though this varies from less than 1 per cent up to 35 per cent. The most likely explanation for this is that the lead helps reduce the tension and viscosity of the enamel and thus 'wets' the metal and helps make the bond more easy and secure.

However it is usually the colouring matter which arouses popular interest. "What causes the colour?" is the question most frequently asked. Like the Iron Age, brighter red enamel of this period continues to be due to copper in its reduced state as cuprous oxide but copper in the colloidal state produces a duller red. Green too is caused by copper but in its oxidised state of cupric oxide. Although copper also causes blue it was found that the deeper blues analysed are the result of the use of cobalt. Yellow is seemingly caused by antimony in the form of lead antimonate while calcium antimonate produces white. Black as was expected comes from iron. Orange appears to be either a mixture of red and yellow, which would explain the varying shades noted between these two colours, or else the result of cuprous oxide with a different particle size from that in the red enamels.

It is surprising to find that blue is the most common colour to survive followed closely by red. Applied less frequently were yellow and green and on some few dozen occasions only, orange, black and white. There proves to be much variation in shade both within one colour and between close neighbours as for instance blue, green and white. Again it needs to be stated that much care is necessary in describing decayed enamels.

The question of opacifiers is not as satisfactory as one would wish. The crystalline cuprous oxide of the reds renders them opaque and likewise the calcium antimonate and lead antimonate act as both colourant and opacifier in the white and yellow enamels respectively. The most likely opacifier for enamel of this period however remains calcium antimonate.

Various other elements are present in the enamels tested. Where an element does not perform a specific function it is usually present in small amounts and may be regarded as an impurity e.g. zinc. The most likely source of these impurities is the sand which contains much more than the desired silica, but various elements could also have been introduced by the use of cullet or from the crucibles or even interaction with the metal base.

It is apparent that the most usual base is a copper alloy. Despite some use of iron on the Continent during the Iron Age this metal was not a medium employed by the enamellers of Britain. Gold appears only once and silver on rare occasions, generally in the post-Roman period. What is generally termed 'bronze' is, then, universal as the base to which the enamel has been applied. Tylecote had previously noted the use of lead in cast bronzes and the widespread introduction of zinc in the Romano-British period. More recently Craddock has pronounced small 'bronzes' to be in fact 'brasses' because of the zinc content. Brass however is said not to be totally desirable for enamelling since too great an amount of zinc will burn off thus affecting the colour. Nearly all the enamelled brooches analysed prove to be leaded bronzes quite suitable for casting, the expected method. Zinc however is absent from a number of the objects subjected to analysis and on these occasions the tin content is generally high at about 10 per cent. Where zinc is present it averages 7.5 per cent and rarely exceeds 10 per cent thus giving an alloy which is best termed gilding metal rather than brass; nevertheless a number do have over 10 per cent of zinc; gilding metal is interestingly highly recommended for enamel-working by modern text books. Another effect of the zinc might have been to produce a more golden appearance in the finished article. The main metal base used for enamelling in the Romano-British period does not therefore appear to follow the normal pattern for small bronzes and while displaying much variation shows no mean knowledge on the part of the craftsmen.

The next problem for the enameller was to make these two separate parts, the vitreous substance and the metal base, come together and adhere. Heat is by definition the instrument for this. The heat came either from a proper or make-shift furnace or was played directly on to the object. A large proportion of the Iron Age red enamel is in the form of studs probably shaped and set in place after having been softened and in some cases further held by an adhesive; but the suggestion by Hughes that the simple inlays of the Iron Age were also emplaced while in the soft state needs to be questioned. He believed that the difficulty of producing red only in reducing conditions necessitated such a method but this is perhaps to underestimate the skill of the enamellers. There is no reason to doubt that Romano-British simple enamels were inlaid either in the form of powder or paste. However the straight edges clearly visible between the different colours of the complex enamels shows that these - except where they act as a ground or bed - were set in position in the form of plaques or slices cut from small rods or canes of enamel. The millefiori too has been applied as slices taken from rods.

There is a great variety of methods of placing the enamel on Iron Age objects. These are best described as uplay and inlay. With the former, studs of red enamel have been set, generally in circular depressions, with or without the aid of central rivets or else in hollow studs or frames. Otherwise it has been placed in cells gouged out of the object's upper surface in what is now commonly known as the champleve technique. In spots, curving cells and geometric cells there are from one to three colours. Despite the great variety of the enamelwork of the Romano-British era all of it is strictly speaking in the champleve technique. Yet this term is patently inadequate to deal with a single coloured plain enamel on the one hand and a millefiori enamel on the other. A detailed examination of the latter material has shown that such does indeed constitute an enamel. A new system has therefore been devised to deal with the enamels based on the manner in which the enamel has been placed in the cell. The basic type has been termed a _simple_ enamel whereby only one colour of enamel has been applied to a cell: the one object can have one, two or three different colours applied simply. There are however other cases where two or more coloured enamels appear in the one cell without any metal divide. These are now called _complex_ enamels. Complex enamels can take the form of different coloured plaques laid side by side or else a ground of one colour into which are laid spots (sometimes with a further ring or rings) of another colour. A sub-group within the complex variety is found consisting of a sulphur yellow ground, the surface of which is liberally sprinkled with small irregular pieces of brown enamel. The third major group consists of _millefiori_ enamels. These are of two kinds: _straight millefiori enamels_ whereby the plaques of millefiori glass have been laid directly on to the base of the cell and fired as an enamel and _floating millefiori_ enamels in which a bed of red (very rarely blue) enamel has been put down first and on top of this have been laid the millefiori slices - on heating the red enamel has become in effect an adhesive for the millefiori. Generally speaking on the Roman-age objects the millefiori completely fills the cell and the red beneath is not visible. In the post-Roman period the millefiori is usually clearly floating in a 'sea' of red enamel. Often in the past borders of red to the millefiori square (i.e. forming part of the pattern) have been confused with beds of red enamel and care needs to be taken over this.

Finally there is possibly another group wherein a layer of blue (glass or enamel) but also occasionally black floats as a thin layer on top of a bed of red enamel. This basic technique has been used with the blue glass spots of the Iron Age and may have continued in use in the Roman period. The best evidence for the method is the black on the sub-group 11 of the disc brooches where microscopic examination has revealed a thin red line around the edge. Yet it could be argued that though this is red cuprous exide it is not in fact a deliberately laid bed but the highly desired formation of cuprous oxide at the interface between the enamel and the metal (Bateson and Hedges, 1975, p. 186). There is however no clear case of this and with blue having been used far more than has hitherto been appreciated it seems sensible to accept the majority as being wholly blue enamels. The idea of continually shaping blue glass and using a red enamel adhesive really seems to be creating an unnecessary difficulty though it would perhaps be unwise to say such a method was not used in some cases at least.

The majority of post-Roman enamels are simple ones with a small proportion of floating millefiori. Despite the existence of garnet cloisonne work cloisonne enamels are unknown from the period covered.

These types of enamelwork all survive from Roman Britain but it is necessary to consider if they were in fact produced by the Romano-British enameller. Simple enamels are found in abundance on such familiar Romano-British objects as the group R and S brooches, dragonesque brooches, button-and-loop fasteners and the miniature stools. Colours range form one to three, rarely more; most frequently two as red and blue. Complex enamels occur less often on British products but were obviously in use. There are some head stud brooches so treated and the sub-type e of the group M brooches as well as the RIVa brooches are decorated in the complex fashion. The oval and medallion plate brooches, though few in number, are an important Romano-British type displaying this technique. These latter groups belong to the first variety of complex enamels with different coloured enamels laid side by side as plaques or squarish slices. For examples of the second variety with a basic ground of one colour into which have been inserted spots of one or more other colours we have to turn to those types of animal plate brooch such as the dog, hare, rabbit and sandal or the seal box lids or finger-rings and other 'miscellaneous' objects and brooches. It has been suggested that more of these have a British origin than is normally accepted but it must also be said for complex enamels that the best examples of this technique are to be seen on objects of Continental origin, particularly the Rhineland brooches and exports of these found in Pannonia. It may be said that the Romano-British enamellers did produce complex enamels but more often of the first variety then the second. Significently the best examples of the technique are to be found on the Continent. There is too the small group of complex enamels consisting of a sulphur yellow ground with brown flecks and for these it is probably best to seek an origin outside Britain, probably the Rhineland; the objects are too few and not sufficiently British to say otherwise.

However perhaps the most interesting question is did the Romano-British enamel-workers produce millefiori enamels? There are over one hundred objects bearing millefiori enamel surviving from Roman Britain. These mainly consist of circular studs, mounts, belt-buckle plates, brooches usually of disc type often with upstanding studs and lugs on the edge and a small number of miscellaneous pieces such as the occasional seal box lid in addition to a small number of the hanging-bowl escutcheons. More of these are obviously of Continental origin (e.g. the disc brooches just noted) than British though the majority are of types which could have a common origin. Yet there can be no such doubt with the Kilpheder brooch which is thus strong evidence for the manufacture of millefiori enamel in Britain itself - and a second such brooch has recently come to light, again in Scotland. These two objects provide the crucial evidence required. Due to the millefiori having disintegrated on the Kilpheder brooch we have the benefit of seeing the real technique used, that is to say the millefiori held in place by a hidden layer of red enamel. We have already concluded that such a method was used and in view of the very similar way of holding the Iron Age blue glass spots in place could be taken as further evidence of a continuation of the ploy into the Romano-British period. In addition examples of floating millefiori are rare from the Continent though there is the piece in the Vatican Museum as well as a brooch from Belgium (Amand, 1972, plate XII). Floating millefiori is more common on the hanging-bowl escutcheons and is more crude (or less proficient) than the Roman-age examples. There is too increasing evidence that some of the hanging bowls were made in Britain (e.g. the Craig Phadric mould for an escutcheon).

There is little doubt that the rectangular mounts with a knob on top, the belt buckle plates of groups 2 and 3 and the lugged brooches are imports into Britain (see Sellye, 1939, plates XVI, XIV and V respectively). Yet we should not over-estimate the extent of such imports to the province. Certain objects are common to both such as the large circular studs and indeed one might argue are more common in Britain; not many come from Gaul or Pannonia. The plain disc brooches (sub-group 1) with only one cell completely filled with millefiori could have been manufactured equally in Britain or Gaul. It seems possible that many may have been produced in the former. Also there are the cow-hide shaped mounts from Leicester, Eden Hall and York and certain seal box lids with millefiori which appear to be absent outside Britain.

It therefore seems feasible to argue strongly for the manufacture of millefiori enamelwork in Britain. However it should be clear that not too much is being claimed in this for the amount produced must have been very small and constitutes only a tiny proportion of Romano-British enamelwork. Again the technique is more common on the Continent in both Belgium and the Rhineland. One might wonder if floating millefiori is a British adaptation of the straight millefiori technique of the Continent but there is little to push this beyond speculation and to do so may be an injustice to the skill of the Romano-British enamel-worker.

This may be an appropriate point to consider whether the raw material (the vitreous substance) was manufactured in Britain or not. Hughes (1972, pp. 102-105) has argued that the Iron Age red enamel was made in a few centres in the Mediterranean and then exported to Britain in 'cakes' or slabs where it was broken up and applied to the objects by the enamellers. Yet the only 'cake' found in western Europe is the unstratified, undated and amorphous stray lump from Tara. It is certainly no cake in the Mediterranean ancient glass tradition (Charleston, 1963, passim). The absence of cakes may be noted from Bibracte and the 'big chunk' of purple glass from Manching remains an open question as to origin (Kramer, 1960, p. 167)[113]. The variation in ingredients (within the basic formula) has been pointed out by Hughes and in particular the Fayum terret enamel is quite off the norm. There is not really enough evidence available to be sure of a Mediterranean origin for such material[114]. This is not to say the red enamel was not manufactured in specialised centres as Newton (1971) as argued re glass but there is no reason to suppose that these should be sought outside Britain. The use of cullet however is an unknown factor and the Traprain Law bangle made from the glass of the Roman pillar-moulded bowl is a good example of its use in a related craft. Certainly Iron Age Britain could have produced the basic glass and the yellow, blue and white found on glass beads and bangles were probably manufactured here as well, even if in a few centres. Two points are noteworthy in this respect: the paucity of red in such articles and the use and availability of blue and yellow[115]. It is thus quite feasible for the enamels of Roman Britain to have been made in the province . Only black is a new colour and this is rarely found. The glassworks at Caistor-by-Norwich, Wroxeter and Wilderspool show that the ability was present though this does not necessarily mean that a connection with enamel-working exists. As regards the millefiori glass there is really no evidence that this was made in Britain and it probably does have a foreign origin. However the advent of glass-blowing in the first century dealt a blow to millefiori glass vessels and one wonders about the continuity of its manufacture. It may of course have dropped to a small scale, exported as a raw material.

Let us now move on to an attempt to date the various techniques. It seems that red enamel was in use by the early years of the second century B.C. There is the small module on the Trevone brooch and the studs of the Maiden Castle brooch. These however represent use on a small scale and it is only in the first century that more objects appear with enamel such as the Bugthorpe discs, the Thames helmet or involute brooches. It is probably only late in the first century B.C. and early in the first century A.D. that there is the great upsurge in enamelwork and probably not until the latter before blue and red together begin to appear. It is exceedingly difficult to date these objects either objectively or from expert opinion. However we should beware of a too easy equating of technical advance with chronological progress. Certainly studs of enamel held in place by rivets such as on the Bugthorpe mounts are early but the incised rivet heads on the North of England helmet are on an object generally dated to the second half of the first century A.D. and the framework studs from Hertford Heath come from a burial assigned to about A.D. 40, though in both cases the objects could be earlier. There is the recurring problem of distinguishing between the dates of manufacture and deposition. For the rise of champleve or simple enamel in spots or larger cells we probably do not need to go very far back into the first century B.C. and the large number of hoards seemingly deposited in the period A.D. 50-70 show that red, blue and yellow in small geometric cells had been in use for some decades at least. Iron Age types persist, especially in the north, well into the second century A.D.

Turning to the Romano-British objects there is little doubt that some of these were being made in the latter half of the first century though mainly those with simple enamels which include red, blue, yellow and possibly green. There are the objects from the Lunt, the group R brooches from Nijmegen, the dragonesque brooch in red, blue and yellow from a Flavian layer in Carlisle, the domed plate brooch recovered at Wroxeter with two coins of Vespasian and the mount from Inchtuthill. There is a head stud brooch from London associated with a coin of Nerva and enamelled in a complex fashion. More important perhaps is the brooch with millefiori enamel, probably an import, found in Saint Albans with a coin of Domitian and Flavian pottery. It is however in the second century that there is a dramatic rise in the amount of enamelwork though the number of closely dated objects is not so great as many would make out - the 'second century' appears at times to take on the garb of 'religious significance' too readily. Simple enamels continue to be made in the popular colours but we can be more certain about the use of white, black and orange. Complex enamels of both kinds are now quite common. Blue and white pieces are found on the group RIVa brooch from Silchester deposited no later than 120 and there are also the smaller complex pieces on the sub-group e of the M brooches of this time. An animal plate brooch in the shape of a dog with a blue ground and red spots comes from the Walbrook streambed as does a casket handle also bearing a complex enamel. While the majority of enamels from Traprain Law are probably of this period they all contain only simple enamel. Newstead however yielded the complete range of techniques on a variety of objects though fewer of these are likely to be of British manufacture.

Millefiori enamel is found on two disc brooches, the first from the Walbrook streambed and the other from a layer dated before 150 at Wroxeter. The millefiori cow-hide shaped mount from Eden Hall was found with coins dated no later than 161. Roman Britain was therefore familiar with, and probably producing, all the techniques examined by the middle of the second century. All too often though we are led to believe that such activity ceased soon after[116]. Yet in examining the various objects in the early chapters we have seen that there are reasons for believing some of these to be later. Even allowing for loss sometime after manufacture it is hard to accept these as 'survivals'. We must try too, to forget the widely held view that the Villa D'Anthee supplied much of the enamelwork found in Britain and that enamelwork became scarce after the villa had been destroyed in the middle of the third century. Third century finds from Britain are less frequent but one can cite the hare-shaped brooch from the Jewry Wall site in Leicester, the disc brooch from a late third century layer at Margidunum or the equally late layer at Richborough which yielded a millefiori disc brooch, and the same site has yielded an oval and medallion plate brooch. We have already seen that this type of brooch is late. Fourth-century finds have been made at Lowbury Hill and Silchester for similar brooches and there is the late date for the horse and rider brooches too, e.g. those from Hockwold. One cannot but notice the close similarity between the heads of the riders and the heads of the medallions suggesting contemporanity and a common source. Brigstock has provided a late miniature stool. There are too the millefiori stud , the belt-buckle plate and disc brooch all from fourth-century contexts at Chesterholm, Caernarvon and Caerwent respectively. While some of the material from Hockwold and Campsfield is undoubtedly earlier, one wonders if more is not later, especially the disc brooches. It has been argued (Hogg, 1951, p. 208) that there was a change in brooch style in Britain at the end of the second century but this should not be equated with the end of brooch production and Clarke (1951, p. 199) has successfully shown that bronze-working continued beyond this time. It seems clear that enamel-working did so also.

Where and whence did these techniques originate, are they indigenous inventions or imports from abroad? There seems little doubt that the idea of red enamel was introduced from the Continent about 200 B.C. Whatever the connection with coral[117], enamel was known long before this on such objects as the Basse-Yutz flagons and Amfreville helmet and during the third century on a series of torcs from South Germany and Switzerland. Somewhere then in western Europe lies the origin of early British enamels. However these are sporadic until the flowering of the 'Opus Britannicum' late in the first century B.C. This has often been attributed to the Belgic invasions (e.g. Fox, 1958, p. 142) yet the Continental background does not seem appropriate for such a wealth of material since the main products there are the rivets with incised heads (Dechelette, 1927, p. 476, figure 404) or a very small number of such pieces as the Nanterre terret (Henry, 1933, p. 75). Secondly it happens too late to coincide with the Belgic invasions and thirdly one of the most prolific areas of the craft north of the Thames lies in Caesar's 'Interior Pars' inhabited by native stock as opposed to Belgic peoples (Harding, 1974, p. 224) although the Belgae may have had a greater say in the production centred on the Severn. Probably also an indigenous idea is, the substitution of red alone with a combination of red, blue and yellow. Not only is progression from red, to red plus yellow and blue spots to finally geometric cells of these three colours discernible in Britain but simple enamels in more than one colour are rare on the Continent, e.g. on the Buisonnet terret (Henry, 1933, p. 106). Leeds (1933, p. 103) thought these coloured enamels owed their origin to Roman brooches but we have seen the technique firmly established before A.D. 60 at which time the imported Roman brooches were almost wholly unenamelled. One might also point to the blue and yellow used on British glass beads and bangles and suggest it is but a short step to applying these colours to enamelwork.

The more immediate impact of the Roman invasion is to be seen in the objects produced and their style rather than technique. The geometric style of cell on brooches replaces the curving cells of the terrets and other paraphernalia of the chieftain. Stanwick however reminds us that this was not a quick process and such objects continued to be made in the north well into the second century as the Middlebie hoard shows. Imports of enamelwork are really what one wants to see when talking about influences yet it must be emphasised that these do not appear to occur in sufficient numbers for some time. We can certainly point to some imported first century enamels: the pair of dagger sheaths from Hod Hill, the lugged disc brooch from Fishbourne (pre-75 deposit), the disc brooch with stud and millefiori from Richborough, the mid-first century seal box lid from Saint Albans and another from a Flavian context at Nottingham, but there are few others. Yet there are over sixty finds of disc brooches with various lugs etc., which for the majority betray a foreign origin. In addition there are numerous lozenge-shaped brooches of one sort or another and over one hundred plate brooches of miscellaneous non-British shapes - Nornour provides some excellent examples. Here we can point to possible prototypes yet the majority, like our own Romano-British brooches, date to the second century. However it is likely from such objects as these that the Romano-British enameller learnt about complex and millefiori enamels and added these techniques to his repertoire. Here too black, white and orange enamels were seen and copied. Also in this century the repertoire of types was extended after seeing such Roman objects as seal box lids, mounts, belt-buckle plates and paterae. The new market of soldiers and the prosperity engendered by the invasion were no doubt responsible for the much greater output of the enamel-working industry which might now be so-called rather than a craft.

Yet if the British enamellers gained much from outside influences it is perhaps necessary to note that objects and possibly ideas were not all one way. British and Romano-British enamelled products did reach the Continent. In the Iron Age one can point to the Leg Piekarski bowl, the Nijmegen mirror and the Fayum and Eauze terrets. Later as Wild, Feacham and Hull have shown button-and-loop fasteners, dragonesque brooches and trumpet brooches were carried abroad. We have noted the Amiens patera and the fragment from Spain now in the Victoria and Albert Museum. There are too the trumpet brooches found in Germany (Collingwood, 1930b, p. 50)[118]. A dolphin brooch (Lambay Island) and a group M brooch (near Bantry) reached Ireland. However the part played by soldiers being transferred to other parts of the Empire or indeed British conscripts or volunteers serving overseas could account for many of these finds and enamelwork can hardly be claimed as having been a significant export.

Now is perhaps the time to try and define which areas of Britain were producing enamelwork. It is not really until the first century A.D. that we have enough objects to try to determine where they were made. To a large extent the main areas where the finds have been retrieved point to areas of manufacture but this obviously is generalised and care must be taken with the internal trade in such objects. We have already examined the possible sites of manufacture. Nornour has been discarded and no site of the stature of Bibracte is known for the Iron Age or Roman period in Britain. Yet it is fairly certain that this activity took place on a dozen sites such as Traprain Law, Colchester, Wilderspool and Charterhouse. Iron Age objects, particularly the hoards, show that much enamelling took place in the south-east probably centred in Norfolk and Suffolk on the Iceni but extending south to the Thames and its estuary. A second area can be located in the south-west covering north Somerset and Gloucestershire and South Wales across the Severn estuary. This might be one area or two and Belgic influence may be strong in the former. Finally we have the north - really a difficult area to define but including northern England and Scotland. The first two were producing simple enamels in red, blue and yellow but to judge by Stanwick the north was at this time somewhat retarded though it continued later in a less restricted atmosphere.

The group M brooches and the group RIVa (closely related to sub-type e of group M) show that the south-west continued making enamelwork in the Roman period into the second century and, although the techniques here included complex enamels, the work is at a generally poor level with small plain cells many of which have lost their enamel. Charterhouse may have been a production centre and it would appear that the craft died out in the second century. The lead model of a penannular brooch from Dinas Powys hardly proves survival and may show later Irish connections. The south-east, despite the upheaval of the Boudiccan revolt, seems to have flourished - or regained momentum - in the second century and throughout the third and fourth. The range of objects is extensive: one suspects the manufacture here of some at least of the disc and animal plate brooches, the domed plate brooches, mounts, vessels, miniature stools and oval and medallion plate brooches. The latter two along with the horse and rider brooches show late production. Techniques certainly include simple and complex enamels and probably millefiori as well. While red and blue are frequent here one finds the wider range of colours introduced by the Romans. No sites are alas known in Icenian territory but the glassworks at Caistor-by-Norwich may have some significance.

In the north enamelwork has suddenly advanced from the rather simple red of Stanwick in the first century to the brooches and other objects of the second. Is this due to movement from the south-west (where enamelling certainly declines) or the south-east (whence an Icenian might have been glad to escape the heavy hand of Rome) or is it merely copying objects used at Roman forts? In any event R, S and Q brooches, button-and-loop fasteners, armlets and belt mounts make their appearance. Complex enamels appear only in the head stud brooches and there is the millefiori on the group S brooch from Kilpheder. Leeds has pointed out the love of red and yellow in the north as seen on the terminals of the massive armlets or the York belt-plate but there is also blue as on the unfinished mount from Traprain. Traprain Law is an obvious manufacturing site and there may be others. It is difficult to know how late this area made enamel or rather to point to objects - such as the Chesterholm millefiori stud - which bridge the gap with the post-Roman objects which seem to have been made in Scotland such as the Pictish pieces or some of the escutchesons for which Craig Phadric provides evidence. However we must be careful in our desire to create schools and areas not to over-simplify matters. Wilderspool does not readily fit into these three areas and another Nornour could unexpectedly disturb the patterns.

Another aspect which we might deal briefly with is the set-up of the industry though this is misty ground. Was the Romano-British enameller established on an industrial basis such as at Bibracte or was he working alone on a much smaller scale?[119]. Wilderspool was an industrial set-up but May was perhaps over enthusiastic in ascribing too much to the enamel-working side of the manufactories there. Yet the large number of head stud, group R and S brooches or the vessels suggest industrialisation on some scale. The 'Wall souvenirs' are hardly the products of personal enterprise. On the other hand the Traprain Law workshop probably never functioned on a large scale and the enamel on the group M brooches does not suggest anything other than work at a low level. Indeed most of the sites listed as possible manufactories seem to imply workshops of limited scope. While no craftsmen of the calibre of some of the enamellers of the Iron Age seem to have survived, the experiments have shown that the enamelwork of the Roman period is not to be despised. What also needs consideration is whether the enameller was solely such or whether he was a bronze-worker who also applied enamel decoration to his products or perhaps a glass-worker who turned his craft to the decoration of bronzes. We have seen that the majority of the cells of the Roman period are cast as part of the object, not gouged separately as in the Iron Age. This could mean either the bronze-worker made the objects to order or completed the enamelling himself. Equally it is feasible for a small scale glass-worker to be an enameller too, for the materials are very similar. We have seen there is sophistication involved in the production of both the metal base and the actual enamel but we should never underestimate the craftsmen and the use of cullet or scrap, so ably proved to have been used at Traprain Law, is an unknown factor in the equation. It is worth noting that the lead model of a brooch at Wilderspool shows bronze-working side by side with glass-working and enamelling. Likewise in one of the areas most prolific in enamelwork we find the Caistor-by-Norwich glassworks. Traprain Law too shows signs of metallurgy as well as glass and enamel-making. On the whole the craft was probably carried out on a small scale by workers skilled in or in close contact with glass and metal manufacture. However Wilderspool or indeed the size of production and spread of some of the groups of finds warns against this being the whole picture. We have seen there was a dramatic change of market from Iron Age chieftains wanting small numbers of specialised objects to a much larger market dominated for a long period by the garrisons. However an interesting feature is the religious side of find-sites. It now seems that Nornour is such and in addition one can point to Woodeaton, Brigstock, Hookwold and Farley Heath where numerous enamelled objects are associated with religious sites.

It is perhaps now time to see if we can construct the story of enamel-working in Roman Britain. This is a story which had its origins over two hundred years before the invasion of Claudius; these early and infrequent red enamels developed into a flourishing craft for which Iron Age Britain was well known and had reached its zenith before the Romans arrived for their permanent stay. The red enamel was applied in a great and ingenious variety of uplay methods but champleve, or preferably simple inlay enamel, came to hold sway. It was applied to harness pieces and weaponry, at first in red, then with blue and yellow, at first in the wild whirls of the Celtic style, then in more sober squares and triangles. The part played by the Balgae in this seems less than has hitherto been considered. By the middle of the first century three areas seem to be producing the bulk of the material: the South-east and South-west (both as defined) and the North. The immediate impact of the invasion was to affect the market or patronage of the chieftains and their warriors for such objects but this was replaced by the much larger market of the army, demanding however a different sort of object[120]. The idea that in the first century Britain was flooded by multi-coloured Roman enamels which greatly influenced the native enamellers does not hold good.

Many of the Romano-British types were being made in the first century such as the head stud brooches, group R and dragonesque brooches. By the second century group S and M brooches, disc brooches, animal plate brooches, vessels, button-and-loop fasteners, seal box lids and a whole range of other articles were being manufactured. However not only did the Romans affect the type and number of objects produced but they introduced a greater range of colours (not widely used though) and the more sophisticated techniques of complex and millefiori enamelling into Britain. These however never attained the proportion of output that simple enamels possessed and remained less popular then on the Continent. These three techniques of simple, complex and millefiori enamels adequately cover the methods found and provide a better scheme for properly describing Romano-British enamels than mere 'champleve'. We have seen that the enamel or vitreous substance itself is similar to basic contemporary glass but has the addition of lead; blue which is found on more objects than any other shade is extensively coloured by cobalt, antimony plays a large part in opacity. The metal is almost always a copper alloy but is divided between a leaded bronze and a gilding metal which has zinc, unlike non-enamelled bronzes which recent studies have shown to be mainly 'brasses'. The so-called use of 'keying' in the form of cross-hatching is surprisingly almost non-existant but in fact is unnecessary in the majority of small cells; however several objects do display toothed or serrated cell edges for this purpose. Areas of production in the second century continue as before but then the south-west seems to fade from the picture. The north extended its repertoire greatly in the late first century and this may have been due to movement from the latter area or the south-east which had a strong tradition of enamel-working. Quite a large number of sites can be earmarked as possible places of manufacture e.g. Traprain Law. It has been shown that enamelling did not come to a halt in the third century but seems to have continued right up to the end of Roman Britain as exemplified by the oval and medallion plate brooches. This provides the missing link with post-Roman enamelling but the products of the latter are in fact much fewer in number than is often implied and show much less variety in technique. They indeed provide the tail end of our story up to about A.D. 700 and do not represent a phoenix-like rising of Celtic traditions.

NOTES

1. For example "from section Q came a fragment of an armlet (fig. 13, no. 17) richly decorated and completely covered with enamel except on the under side" (Cree, 1923, p. 201).

2. The composition of niello is said to vary considerably but "was normally prepared by fusing sulphur with silver and copper, although lead and antimony might also be added" (Hodges, 1964, p. 79). Niello can occasionally be mistaken for enamel especially when it is in a decayed state.

3. There seems little point in listing and illustrating every enamelled brooch since these all form part of the late M.R. Hull's forthcoming corpus of brooches from Roman Britain.

4. The search for material was directed two ways, towards the museums on the one hand and the published sources on the other. Apart from personal knowledge and the advice of my supervisor, the two major lists of British museums were consulted to see what museums exist and what types of collections they house. These were The Museums Calendar (including a Directory of Museums and Art Galleries of the British Isles) published by the Museums Association (London, 1972 edition) and Museums & Art Galleries in Great Britain and Ireland 1972/73 (ABC Travel Guide, London). There follows an alphabetical list, by location, of the museums visited and where all relevant enamelled objects were examined personally: Alnwick, Alnwick Castle; Bath, The Roman Baths and Museum; Bristol, the City Museum; Caerleon, The Legionary Museum; Cambridge, The Fitzwilliam Museum and The University Museum of Archaeology and Ethnology; Canterbury, The Royal Museum; Cardiff, The National Museum of Wales; Carlisle, The Museum and Art Gallery, Tullie House; Chester, The Grosvenor Museum; Chesters, Roman Fort and Museum; Cirencester, Corinium Roman Museum; Colchester, The Colchester and Essex Museum, The Castle; Corbridge, Corstopitum Roman Site and Museum; Edinburgh, The National Museum of Antiquities of Scotland; Leicester, The Jewry Wall Museum; London, The British Museum (Departments of Prehistoric and Romano-British Antiquities and Medieval and Later Antiquities), The Guildhall and London Museums now amalgamated to form The Museum of London, and The Victoria and Albert Museum; Newcastle-Upon-Tyne, The University, Museum of Antiquities; Newport (Mon.), Museum and Art Gallery; Norwich, The Castle Museum; Nottingham, City Art Gallery and Museum, The Castle; Oxford, The Ashmolean Museum; Reading, The Museum and Art Gallery; Richborough, Richborough Castle; Saint Albans, Verulamium Museum and Hypocaust; Tamworth, The Castle Museum; Taunton, Somerset County Museum; Winchester, The City Museum; York, The Yorkshire Museum. The help received at these museums is gratefully acknowledged. For a number of other museums time or circumstances permitted only an examination of the material on display.

5. The search of published sources concentrated on the journals and produced a wealth of material. There follows a list of those journals which were systematically checked from their first issue through to that for 1980 where this was available at the time of writing: Antiquaries Journal, Archaeologia, Archaeological Journal, Britannia, Medieval Archaeology, Proceedings Society Antiquaries Scotland, Antiquity, Proceedings Prehistoric Society.

6. There is not much to be said on the literary aspects of enamel-working. While no search was made of the classical writers it seems reasonable to conclude from what has been written on the subject in modern times that such a search would prove fruitless. This note therefore concentrates on the well known and frequently quoted statement of Philostratus. Little is usually said beyond giving the reference and the author's birthplace.

   Use has been made here of the Loeb Classical Library edition of Philostratus Imagines with an English translation by Arthur Fairbanks (London, 1960, Heineman). The Philostratus in question is commonly known as 'Philostratus the Elder' who probably wrote the Imagines early in the third century A.D. These Imagines are to be distinguished from those written about A.D. 300 by his grandson 'Philostratus the Younger'. The former describe a series of paintings though how real these are is unknown. The particular passage of interest is to be found in Book I, number 28 entitled 'Hunters' and may be translated as follows:

   > 'They say that the Barbarians by Oceanus pour (these pigments) onto red hot bronze and the colours join together and harden and preserve what has been painted'.

   There is not a great deal that can be said on this. It seems reasonable that 'Oceanus' refers to the Atlantic. Henry (1933, p. 66) had no doubt that the Barbarians were western and not eastern, that it was the Atlantic Ocean and not the Black Sea. The Barbarians are likely to be the Celts of Britain and/or Gaul. Philostratus appears to have realised that enamelling was a native thing and not purely Roman. It is interesting therefore that it should appear in a purely Roman context far removed from the centre of production or if solely on imaginary paintings in the poet's mind that he knew of its existence. The use of enamel on bronze accords with what is known of the enamelled objects and the use of the word probably does not have any significance.

7. Unpublished doctoral thesis.

8. Professor Jope has let me see a copy of what Hull has to say on enamel: it is short and of little help.

9. One example of this is the terret found at Lesser Garth, Glamorganshire (Savory, 1966, pp. 28-33).

10. British Museum reg. no. 1909.5-3.1; see Moore (1973, p. 158) for a list of type 2 terrets.

11. This is generally known as the 'Suffolk Terret' (Leeds, 1933, p. 40) but Piggott (1952, p. 21) has pointed out that probably it is to be identified with that listed from Lakenheath (Clarke, 1939, p. 70). The piece is in the Ashmolean Museum, reg. no. 1927.4615.

12. See Maxwell (1886, pp. 396-398, plate VIII, no. 2) - National Museum of Antiquities of Scotland, reg. no. FA 40.

13. Such as that from Cawston (Clarke, 1939, p. 70 and fig. 12, no. 3). The published drawing of this object shows it with the missing enamel partly restored and with a small hole in the centre of the middle circular (spot) cell giving the impression of a rivet hole though such is unlikely.

14. The example from Colchester (Colchester and Essex Museum) with the red enamel restored under the lips seems to be an incorrect restoration rather than an exception. The object was found in Causton Road in three pieces and may represent two terrets rather than one.

15. Curle and Cree (1922, p. 258, fig. 24).

16. British Museum reg. no. 80.12-14.11.

17. See Curle (1932, p. 365, fig. 49).

18. Brailsford (1975, p. 230) suggests coral or glass and merely lists the two pieces as 'dolphin-shaped objects' but quotes Fox (1958, p. 230) who describes them as ornamental harness mounts. Similar settings occur on the trace hooks from Polden Hill and the surviving studs on these are not enamel but may be coral (Brailsford, 1975, p. 230, pl. XXIIIb).
Brailsford (1975, p. 230) indeed refers to the normal cheek-pieces from Polden Hill as 'toggles'.

19. Ward Perkins (1939, p. 191) suggests a 'coral or enamel' stud for the Hengistbury Head setting while Barker and Megaw (1963, p. 213) say "coral". The similar setting on the Llyn Cerrig Bach bit was believed to have been "(red?) enamel" (Fox, 1946, p. 28). While coral is a possibility enamel seems more likely.

20. This bit is in the National Museum of Antiquities of Scotland reg. no. FA29. Smith (1925, p. 103) mentions only red enamel while Leeds thinks "yellow, not merely red as usually stated" (Leeds 1933, p. 116). At present the enamel has mostly disappeared but there do appear to be some traces of red still. However Leeds is probably right in believing the former existence of a second colour. See also Ward Perkins (1939, p. 183).

21. Quatrelobe is the most common term for the first group though often applied indiscriminately to the cruciform group which is quite different even if related. The terms are from Henry (1933, _passim_). Cruciform is a term not used much outside the latter work.

22. See Fox (1958, p. 127, 52d) who gives the provenance as 'near Cheltenham'; this same object is noted in the Antiquaries Journal (1938, pp. 75-76) as coming from '? near Sudeley Castle' and is now housed in the British Museum. The correct provenance is the Sudeley Castle Estate.

23. See Wild (1970, p. 155, no. 162); the original of this object is in the Art Gallery and Museum, Perth.

24. This object in the Yorkshire Museum (reg. no. H141. 130 (K30)) is supposed to be from Fremington Hagg but doubt has been cast on this (Webster, 1971, p. 120).

25. These are mounts from Brough (Museum and Art Gallery, Carlisle, reg. no. RF230); two from Colchester (Hawkes and Hull, 1947, p. 334, no. 37 & p. 339, no. 21); Iwerne (Hawkes, 1947, pp. 51-52, fig. 8, no. 3, where the description as a belt-link does not seem as good as a piece of harness equipment - Bulleid and Gray, 1911, p. 228); and one from London (Leeds, 1933, p. 54).

26. Stead (1979, pp. 58-59) describes these as shield mounts but more importantly gives them the detailed description they have so long merited. However this is not absolutely clear in relation to the decorative studs and should perhaps be treated with some caution.

27. This well known object(s) (British Museum reg. no. 70.10-13.5) has never been adequately published. The enamel decoration on it (see Fox, 1958, p. 118, fig. 73, 2) is as follows:

Scabbard: On the front are attached five narrow bands - four of similar size, one smaller - with eight and seven small rectangular cells respectively; most of this enamel is decayed but red, yellow and possibly blue are clear; some of it is now a green colour but this is the result of decay rather than being deliberate. Also attached to the front are two rivets - and possibly once three more - the heads of which are flat and have a spot and ring of enamel, now decayed but possibly blue and red.

Chape: on the front, above the lips, are two cast (not the heads of rivetted studs) spot and ring ornaments of yellow and red enamel.

Sword hilt: bottom - row of twelve small cells alternately red and a greenish colour which may be decayed blue; above this are five cast spot and ring ornaments of yellow and red; then there is a central band with cells of red and decayed enamel alternating; at the top is a band of ?sixteen small cells alternately red and a decayed colour, now white.

28. This is in the Yorkshire Museum and is usually described as being from either Thorpe or Rudston(e) - this should be Thorpe, near Rudston. There are seven (plus one now missing) applied studs, the heads of which are inlaid with enamel. This is simple inlay and the use of the word framework by Leeds (1933, p. 44) should not be equated with the use of this term in the present work. See also Stead (1979, pp. 62-63, fig. 21,6 and pl. 9).

29. Simple inlay on the Birdlip, Old Warden and Nijmegen mirrors but also on the last and on that from Holcombe simple studs of red enamel without rivets.

30. The second colour is described by the Wheelers as green but Leeds (1933, p. 93) thinks it rather to be light blue with a green tinge which would seem to be the more likely colour.

31. Fox (1958, p. 77) says there are traces of blue remaining - he also suggests there might once have been white pupils. Fox (op. cit.) further strongly suggests that the heads of the rivets or nails driven through the plain loops at the backs of the heads (to hold them steady) carried bosses of red enamel though there is no evidence for this.

32. Information from Professor Jope.

33. Savory (1966, p. 31) suggests the possibility that the two hollow studs may not go with the solid one which is larger and that two terrets may therefore be represented.

34. This pair were found at Pitkelloney, near Drummond Castle but either place is given as their provenance; both pairs are in the British Museum (reg. nos. 1946.4-2.2 and 38.07-14.3).

35. See Ellis (1829, pp. 285-296). This object was found in the Culbin Sands near Altyre, Morayshire.

36. Stamford Hill is sometimes used instead of Mount Battan; see Fox (1958, p. 14).

37. Reg. no. 1963.4-7.1 - possibly from Dorset and sometimes known as the 'Portland Twin'.

38. The lesser Garth hoard is not included since it contains only the enamelled studs (for a terret) which are dealt with under the terrets.

39. Spratling (1975, p. 206) for find-spot and promised publication.

40. There is also one harness mount or strap union in the British Museum (Davis and Spratling, 1975, pp. 123 and 131, no. 18).

41. Hull's forthcoming corpus with its much greater number of types dealt with in detail will radically alter this situation.

42. See Anderson (1905, pp. 374-376); the two brooches are in the National Museum of Antiquities of Scotland, reg. nos. FA61 and 62.

43. Each found still on approximatley forty specimens, sometimes together.

44. The green areas on the brooch from Faversham (British Museum, reg. no. 1078-70) when analysed were discovered to be decayed

red (Painter and Sax, 1970, p. 167).

45. Nornour would be included here though see chapter 9 on 'sites' for details.

46. There were eight examples of R(i) brooches noted but no enamel is to be seen on any of them - however it is possible that there once was on the heads or bows. On the brooch from Wilderspool (May, 1904, p. 81) there was said to be "traces of black enamel on the front" but this seems unlikely.

47. This is in the Fitzwilliam Museum, Cambridge, reg. no. GR3.1822; there are only a few traces of the blue left in the now white substance on the trumpet but enough to show that the whole was probably once blue.

48. The size of this cell is 15 mms. x 6 mms. and the individual pieces 4 sq. mms. approximately. The colours are described as orange and red (Thompson, 1965, p. 83, fig. 20, 3) but decayed and orange seem safer. The object is in the Municipal Museum and Art Gallery, Warrington.

49. See Boon and Savory (1975, pp. 50-57) for the development and dating of R(i) and R(ii) brooches in relation to the trumpet brooch found at Carmarthen.

50. Throughout this section reference will be made to these numbers 61-68 as (61) etc., and see particularly Collingwood's drawings of the S group variety similarly numbered (<u>op. cit</u>. fig .103).

51. This type is rather like a fly or other insect and is sometimes described as such.

52. This is unlikely to have been one complete slice but rather a number of smaller slices placed together to fill the cell and then fused together by the heat to form one overall pattern.

53. A second similar, but larger, 'Kilpedher type' brooch has recently been recovered during the excavations of the broch at Leckie, Stirlingshire. On this specimen the millefiori is still present both in the disc on the bow as well as in the fantail. Its date of loss is probably early second century A.D. I am grateful to Dr. Euan MacKie, the excavator, for this information. Separate publication of this important find is planned.

54. This number includes those found on the Continent. There may now be added: 1) portion of a type 1 specimen from Chesters (Budge, 1907, p. 387); 2) one found near Whittington Hall, Gloucestershire (Dawkins, 1874, p. 98); 3) a specimen recovered from the East vicus at Watercock (Olivier, 1979, pp. 210-211, no. 14, fig. 84, 14).

55. Although not examined, the chalcedony may in fact be a decayed enamel which has taken on the appearance of chalcedony - we have already seen examples where the enamel has weathered to the

most misleading of appearances. In this case blue may have been the original colour, if enamel it is.

56. All three are now in the Ashmolean Museum, reg. nos. R68, 67 and 71 respectively.

57. There is now a brooch of what appears to be the large rather than the small variety from the Winterton Roman Villa which was found in a fourth century context but can probably be regarded as a survival to, rather than, a product of, that period (Stead, 1976, p. 201, no. 32, fig. 101, 32).

58. The second colour on the Silchester brooch appears to be a dark green though Hunter and Kirk (1952, p. 60) describe it as being black; however they also say the centre of the medallion bears "an indecipherable design" though this is clearly a human bust facing left.

59. Apart from those from Pfunz (Bushe-Fox, 1949, p. 117) the Lowbury Hill specimen has a parallel from Gelb on the Rhine (Atkinson, 1916, p. 35).

60. This late type usually round, sometimes oval, and mostly gilded has a dark-coloured stone setting in the centre but no enamel; some other plate brooches with enamel however also have a (?) stone set in the centre - where this has vanished (and this also applies to the type with the metal cone in the centre noted under miscellaneous disc brooches) the brooch then resembles the plain disc brooch under discussion with, it could be mistakenly said, enamel gone from the centre.

61. These have lately been discussed as dress fasteners again and at the same time Wild's arrangement has been criticised (Kilbride-Jones, 1980b, pp. 159-169).

62. This list was compiled principally from published sources and includes several Continental examples (Wild, 1970, p. 137). There are for instance four enamelled objects of this type in the site museum at Corbridge not included.

63. Classes I and II are basically Iron Age in type and are dealt with in chapter 1. Although class VIIIb is said to be sometimes enamelled (Wild, 1970, p. 143) only two from Britain (Hod Hill) are listed (<u>ibid</u>. p. 154, nos. 128 & 129) and neither bears enamel.

64. Other suggested functions include stands to hold unguent bottles (Franks, 1854, p. 27), altars (Wheeler, 1946, p. 111) and even doll's house furniture (Barnett, 1969, p. 23): more recent discussion tends to confirm their religious function (Green, 1975, pp. 57-59 and Butcher, 1977, p. 51).

65. These are not always recognised as such, e.g. the 'perfume box' (Trollope, 1857, p. 75), especially when only the lid survives since it resembles an ornament or mount of some sort. The group found at Dura Europos is described as being lockets for some

medicinal or magical preparation (Toll, 1949, p. 37).

66. Always on the upper outside surface; the lids are either solid or slightly 'hollow'.

67. Sometimes there are small knobs or projections but the shapes generally conform to the types listed.

68. Information gratefully received from the excavator, Mr. Gavin Simpson.

69. The original is in the museum at Alnwick Castle (Bruce, 1880, pp. 139-141, no. 746); an electrotype copy is in the British Museum.

70. Henry says this is in the Canterbury Museum but there is no trace of such an object in the Royal Museum at Canterbury.

71. A third one has been found during the 1975 excavations in York in a post-Roman pit of the seventh century; this appears to be a stray survival; I am grateful to Mr. R. Hall for this information.

72. An electrotype copy only is in the National Museum of Antiquities of Scotland, reg. no. FA 110 - 1971.55; the original is in the museum at Dunrobin Castle.

73. The enamelled bracelet from Rhayader is dealt with in chapter 1, and Butcher (1977, p. 68, note 5) mentions a fourth-century enamelled bracelet from Frocester.

74. Only the base survives and it is treated here as belonging to a candlestick. A fourth in the Fitzwilliam Museum, Cambridge (Dalton, 1912, p. 100, no. 45) is hardly Romano-British.

75. The best known are the Alfred and Minster Lovell jewels on which the enamel decoration is quite different from anything met with in the present work (see Wilson, 1964, passim and Hinton, 1974, passim).

76. Now see Kilbride-Jones's detailed study of the zoomorphic penannular brooches, particularly as regards dates and attribution (Kilbride-Jones, 1980a). Note too the specimen from Abingdon, Oxfordshire, with red enamel in the terminals; this brooch has a close parallel in one from County Armagh (ibid. p. 93, no. 25).

77. It is difficult to know whether to include the ibex-headed pins here or not. First or fourth century dates are the most commonly accepted but Fowler (1963, pp. 123-125) seems to suggest a sub-Roman date as does Brown (1976, p. 20) in relation to the specimen from Cirencester. The enamel decoration on this latter example is very similar to that found on the ibex-headed pin from the Dundrum Sandhills. The only other enamelled specimens from Britain are those from Newham, Cambridgeshire, and from grave 423 at the Duston Anglo-Saxon cemetery, though

the latter probably pre-dates its sixth century deposition (Fowler, 1963, p. 123, fig. 7,3 and p. 155). These all possess only very simple red or decayed enamel. Kilbride-Jones (1980b, p. 248) follows Stevenson in suggesting a Scottish origin in the fourth century.

78. It is now claimed that this pin is of Irish manufacture, no later than the middle of the third century (Kilbride-Jones, 1980a, p. 22 and 1980b, pp. 212-213).

79. There is an unprovenanced specimen, with three fingers and millefiori enamel (floating) on the head, in Alnwick Castle. However much of the material in this collection is from Ireland and this handpin may therefore possess an Irish find-spot (Bruce, 1880, p. 91, no. 505).

80. Again see Kilbride-Jones's comments on these objects and the handpin from Gaulcross (Kilbride-Jones, 1980b, p. 152).

81. Conservation work being carried out at the time research was being conducted for the present work precluded close examination of the Sutton Hoo material. The discussion on this is therefore mainly based on the published sources.

82. See too Kilbride-Jones on the subject of hanging bowls (Kilbride-Jones, 1980b, pp. 236-257).

83. There is in the Alnwick Castle collection a disc, from a burial on Coquet Island, which could be an escutcheon. It bears a small cross in the centre and has an outer circle of shield- or tongue-shaped cells alternately filled with blue and yellow enamel (Bruce, 1880, p. 74, no. 285, figure). Another possible escutcheon, with millefiori, is recorded from a cist burial on Hallowhill, Saint Andrews, Fife (Webster, 1977, p. 217). There is too half of what appears to be an escutcheon from Aberdour Castle, Fife, now in the National Museum of Antiquities of Scotland (reg. no. FC298); it is in poor condition but appears to have yellow and another decayed colour, possibly red.

84. There is also the clay mould found at Craig Phadrig, Inverness-shire (Stevenson, 1972, pp. 49-51).

85. There also appears to be millefiori on the escutcheon from Baginton (Kilbride-Jones, 1980b, p. 240, fig. 77,5).

86. Note might also be made of a few analyses carried out on the material from the Villa D'Anthee where the colourants used for a blue and a brown rod were found to be copper and manganese respectively; copper too is used to colour the blue powder in which lead again proves to be present (Spitaels, 1970, pp. 240-241). Copper silicate is said to have coloured the blue frit recovered in Colchester (Hawkes and Hull, 1947, p. 345).

87. See Bateson and Hedges, 1975, pp. 177-190 especially pp. 177-181 for a list of the objects, reasons for their selection, methods used and tables of results.

88. The numbers in brackets used here and subsequently refer to the objects so numbered in the paper on the analyses (Bateson and Hedges, op. cit.).

89. For an example of such a process see Bayley, 1978, p. 254.

90. Jope makes an interesting suggestion in this respect (Jope and Wilson, 1957, p. 83). See also an article entitled, "Where is the Tin?" (Charles, 1975, pp. 19-24).

91. These figures are for fine gold, silver etc. and are taken from Maryon (1954, p. 301).

92. Theilade gives 5 per cent, Smythe 13 per cent, Maryon c.16 per cent, and Tylecote up to 20 per cent; all references to works cited.

93. Shaw too says that brass cannot be effectively enamelled and even includes bronze in this category (Shaw, 1972, p. 31) but he modifies this view in suggesting that a copper/zinc alloy with 9-10 per cent zinc yields good effects when handled skillfully (ibid. p. 39) and with 5 per cent zinc gives excellent results (ibid. p. 36).

94. See however the analyses carried out on some of the Gussage All Saints material (Wainwright and Spratling, 1973, p. 127, and Spratling et. al. 1980, p. 280).

95. This is because zinc melts at 428°C and boils at 950°C. The difficulty was overcome by mixing calamine ore ($ZNCO_3$) with copper in reducing conditions in order to produce gilding-metal/brass.

96. The only enamelled objects whose metals have been previously analysed are: one of the British Museum's head stud brooches (Painter and Sax, 1970, pp. 164-165, no. 13) where the body of the brooch is "a type of brass", the stud on the bow essentially copper and the rivet fastening it a true bronze - no actual amounts are given however; the Southacre bracelet was found to be bronze though again no data is supplied (Mottram, 1968, p. 259).

97. "Thus out of the 80 trappings, fibulae and brooches that I have analysed, only five were not of brass...." (Craddock, 1974b, p. 349).

98. This may have been due to a scarcity of copper or to a desire for the greater strength of an alloy.

99. For instance the use of rivets to hold studs of enamel in place or the use of a framework with some of the studs; these are discussed in the next chapter on techniques.

100. A suggested typology of enamelwork is set out in the next chapter.

101. For a discussion of such tools and their use see Lowry et. al. 1971, pp. 167-182.

102. The moulds found at Gussage All Saints are for objects too early to have been enamelled but it is worth noting that the relief decoration appears to have been made on the patterns before casting (Wainwright and Spratling, 1973, p. 122).

103. Some of the Romano-British flat circular studs or mounts have a depression in the centre which might have resulted from the use of a drill or lathe.

104. An unusual method of keying (in that small spikes of metal were punched or pecked up by a pointed tool) is to be seen on a bronze ferrule from the River Bann (Jope, 1951, p. 57).

105. See Bateson and Hedges, 1975, pp. 185-186.

106. Information from Dr. J. Raftery.

107. See also Jones et. al. 1976, p. 240.

108. These specks are sometimes described as being black but it is doubtful if both black and brown were applied and brown seems to be the correct description. What appears to be black is often in reality brown, dark green or even purple (Harding, 1960, p. 127; Guido, 1978, p. 15; Butcher, 1978, p. 218).

109. The question of manufacturing sites is briefly discussed by Butcher (1977, pp. 42-44).

110. It has unfortunately now appeared in the literature as a major enamel-working site (Kilbride-Jones, 1980a, pp. 16, 75 and 1980b, p. 36).

111. See also Biek, 1977, p. 70.

112. Not included in this list is a mould 'from the frontier area' said to be for the manufacture of enamelled studs (Miket, 1974, p. 162). Another mould (unless it happens to be the same one) in stone has come from Gateshead and it is suggested that this may have been intended to produce miniature wheels or wheel-brooches (Green, 1978, p. 60, pl. 49). When examined it did not appear to be relevant to the present study.

Note might also be made of the only rod of millefiori glass to have been found in Scotland from Luce Bay (Cramp, 1970, p. 333) though it may never have been intended for enamelwork. The same can be said for the stump of a rod of distorted millefiori glass recovered during the excavations at Jarrow (Cramp, 1970, p. 330).

113. The fragments of purple glass found at Hengistbury Head are generally interpreted as having been imported from abroad for the manufacture of bangles on the site (Cunliffe, 1978, pp. 42-44).

114. See Spratling (1980, p. 114) who quotes Megaw on a similar conclusion.

115. See Guido, 1978, pp. 16-18.

116. Henry, 1933, p. 123 and Kilbride-Jones, 1980b, p. 16.

117. See Champion, 1976, pp. 29-40 especially page 35 where it is said that it can be no longer accepted that enamel was developed as a substitute for coral.

118. See also A. von Bohme in "Englishe Fibeln Aus Den Kastellen Saalburg Und Zugmantel", Saalburg Jahrbuch, 31, 1974, pp. 5-20.

119. An interesting comparison may be made with that part of the modern enamel-working industry which deals with the enamelling of small objects such as badges. Although quite extensive and concentrated in Birmingham the industry there consists of many small workshops (owned by individual firms), each generally employing fewer than five workers (Smith, 1976, pp. 34-34).

120. See also Webster, 1978, pp. 490-491.

APPENDIX A

SOME EXPERIMENTS

Something should perhaps be said about the experiments carried out to aid the study; and the tools and raw materials used for these. The purpose of the exercise was to gain some practical experience in the craft, to see how easy or difficult the process is and to answer some questions such as how ancient millefiori enamel was created.

All the equipment and materials were purchased from a London firm, 'Crafts Unlimited'. The kiln used is called a Hobby 11 with a firing chamber 160 x 175 x 110 mms. and a maximum temperature of 1200°C. Otherwise the actual tools needed are surprisingly few and simple - many could be improvised. These include a mortar and pestle (not essential when powdered enamels are bought), a spatula for placing the enamel in the cells, a fork and tongs for placing the object in the kiln, asbestos gloves which are useful but not essential, a wire mesh firing support and an abrasive polishing sponge as well as a carborundum stone; an asbestos pad is also very useful.

The objects to be enamelled are described as 'copper' and have the cells already cast - the bases of these are rough. I mainly used simple rectangular blanks with only one cell but also occasionally blanks with more elaborate designs. The enamel comes in the form of powder or lumps and the range available is very extensive as regards colour and opacity. The melting temperature is about 800°C. I also purchased the enamel in small squarish plaques and spots to try to make complex enamels. The most useful product of all is the millefiori which in fact has only become available to hobbyists within the last few years. This comes in the form of slices 2-4 mms. thick cut from a circular rod of up to 8 mms. diameter. It is interesting that the majority of these goods are imported from Germany. Some are shown in figure 14A.

One's first reaction to the process is how easy it really is and despite warnings from textbooks about cleanliness and care over timing etc. there is much more leeway here than expected. In any event one suspects the ancient enamellers in rather poorer circumstances not to have paid too much attention to such things.

The kiln heats up to its maximum temperature in about an hour. It cannot be regulated to an exact temperature and I generally kept it on full heat during use to compensate for heat loss when opening the door to place or remove the object. This loss is appreciable. The enamels therefore need much less time to be fired though this does not noticably affect the result. About a minute is ample time for a simple enamel and two to four minutes to reach the desired stage with a millefiori one. No experiments were attempted with blow-pipes or open hearths.

As regards putting the enamels into the cells it proved much more convenient to use a paste rather than a powder. The latter tends to spread on to the cell walls or where two colours are used to merge unevenly more easily than with a paste. A few drops of water are sufficient to create a paste which is more easy to control - here a spatula is quite essential. It is important to let pastes dry out thoroughly before firing otherwise they tend to 'explode' as the heat rapidly drives off the moisture. Once removed from the heat they cool reasonably rapidly in normal room conditions; nothing remarkable emerged as regards cooling. Generally at this stage the enamel contracts quite a lot and no longer fills the cell completely. A second firing, and even a third sometimes, is required to finish the process entirely. One is supposed to fire different coloured enamels in order, starting with that having the highest melting temperature. However I fired different colours at the same time without noticeable adverse affect. Indeed otherwise one would need a great knowledge of exact firing temperatures and a very refined control over the kiln.

It proved quite easy to make complex enamels. The squarish solid plaques (see figure 14A - middle of bottom row) are assuredly similar to what a Romano-British enameller cut off the end of his rods of enamel. Once placed in the cell careful handling keeps them in place and they fire just like an ordinary enamel. A little longer time is necessary for the modern plaques to fill the cell fully but presumably the ancient enameller had more control over the size of his cells and rods. It would probably be more handy to use a gum to keep the plaques in place. 'Enamelfix' or gum tragacanth are used nowadays; these fire out easily without discolouring or affecting the enamel. To make the second type of complex enamel I first put down a layer of red enamel in powder or paste form and then set the solid spots (similar to the squares above but from smaller rods) into this and fired the two together. This works admirably and seems the easiest and most sensible method of making this kind of enamel.

The experiments with the millefiori worked exceedingly well despite being handicapped by having only circular pieces which were too thick. However it was proved that such a material can be treated simply as an enamel. In fact using a bed of red enamel to hold it in place is a more tedious job which seems less pleasing in the end result. I have already discussed the actual process in detail in chapter 8 and will not repeat it here. However I find it much easier to make the process work when the kiln is hot though when cooler it is merely a question of allowing much more time. Once the pieces have begun to spread it is essential to keep a very close watch on the object, for a matter of five to ten seconds can result in the loss of the patterns altogether. Usually however it is the outer colour or band which spreads first and quickest though indeed this is what one wants to fill the cell and create one overall pattern.

On the whole the experiments were well worthwhile. They not only answered such questions as the possibility of millefiori enamel very satisfactorily but they provided a valuable insight into the workings of the craft and enabled one to look at the Roman-age enamels with greater appreciation.

APPENDIX B

LIST OF ENAMELLED HANGING BOWLS

These are listed in alphabetical order by site with the county added. A number after a site indicates the presence of more than one bowl. The first column indicates what has been enamelled i.e. escutcheon (E), print (P), bands (B), plaques (pq.); the number before the first two denotes the number known; an asterisk after indicates the presence of tinning. The next column deals with enamel: red (R), blue (B), yellow (Y), green (g), decayed (D) or gone entirely (G). The third column gives the technique employed whether simple (Sx1, 2 or 3), complex or millefiori.

Site	Enamelled	Enamel	Technique
Aberdour Castle, Fife	1E*	Y+D	S x 2
Baginton, Warwickshire	4E & 1P	R	S x 1
River Bann, Co. Londonderry	1E	G	S
Barlaston, Staffs.	2E & 1P	R	S & M
Barrington A, Cambs.	1E	G	S
Barrington B, Cambs.	1E*	D	S x 1
Basingstoke, Hants.	3E	D	S
Bedhampton, Hampshire	1E or 1P	-	S
Benniworth, Lincs.	1E	-	S
Benty Grange, Derbyshire	2E*	D(Y)	S x 1
Bekesbourne, Kent	2E & 1P*	R	S x 1
Camerton, Somerset	3E	Y	S x 1
Chalton, Hampshire	1E*	D	S
Chesterton-on-Fossway, Warwicks.	4E & 1P	R	S
Coquet Island (?)	1E	B+Y	S x 2 & C
Dover, Kent (3)	5E & 2P & B	-	S
Faversham, Kent (4)	3E	D/G	S x 1
	3E & B*	D	S
	1E & 1P	D	S x 1
	2P*	D	S x 1
Greenwich, Kent	3E*	D	S x 1
Grindlow, Derbyshire	2E	-	-
Hildersham, Cambs.	3E & B & pq	D	S
Hitchin, Herts	2E*	R+Y	C
Keythorpe Common, Leics.	3E	-	S
Kingston Down, Kent (2)	3E & 1P	R	S x 1
Loveden Hill, Lincs. (3)	3E & 2P	D	S

	3E	D	S
	2E	D	S
Lowbury Hill, Berks.	3E & 2P*	R	S x 1
Lullingstone, Kent	4E & B & pq.	R	S x 1
Market Rasen, Lincs.	2E	R	S x 1
Middleton Moor, Derbyshire	1E	R+Y	S x 2
Mildenhall, Suffolk	3E & 1P & B	R	S
Morden, Surrey	1E	D	S
Needham Market, Suffolk	2E & 1P & B	?	S
Unknown, Northumberland	3E & 1P	-	M
Oving, Bucks.	1E	R+Y	S x 2
near Oxford	1E	R+B+Y	S x 3 & C
Saint Ninian's Isle	1P	R	S x 1
Scunthorpe, Lincs.	2E & 2P	R	S & M
Stoke Golding, Leics. (2)	2E	R	S
Sutton Hoo, Suffolk (3)	3E & 2P & pq.	R+ B	M
	2P	R	S x 1
	3E	R	S x 1 (+ M)
Thornham, Norfolk	1E	D	S
Tottenhoe, Beds.	1E	R+Y	S x 2
Winchester, Hants.	3E & 2P*	R	S x 1
Unknown, Westmorland	1P*	D	S
White Horse Hill, Oxon.	1E	R+Y	S x 2
Whitby, Yorkshire (3)	2E & 1P & B	R+D	S x 2
Willmington, Sussex	1E	D	S x 1
Willoughton, Lincs.	1E	R	S x 1
Winchester, Hants.	3E*+1P	R	S x 1
(River) Witham, Lincs.	4E	-	M
Unknown (Victoria & Albert)	1E	D	S x 2
Unknown (British Museum)	1E	D	S

APPENDIX C

THE MILLEFIORI PATTERNS

This appendix contains a list of the patterns of the slices of millefiori glass used in the enamelwork under examination. The first column gives the pattern number and the second a description of it. The third column gives the number of times the pattern has been noted.

Type A - basic pattern of nine squares (figure 12)

A1	five white and four blue	3
A2	five white and four red	4
A3	five red and four yellow	5
A4	five blue and four white	1
A5	five yellow and four red	1
A6	four blue and four yellow with red centres	1
A7	five white and four blue plus red border	21
A8	five blue and four yellow plus red border	1
A9	five blue and four white plus red border	6
A10	five blue and four white plus blue border	1
A11	five white and four blue plus blue border	3
A12	five white and four red plus blue border	1
A13	four blue and four white with red centre plus blue border	2
A14	four white and four blue with red centre plus white, plus blue border	1

Type B - basic pattern of twenty-five squares (figure 12)

B1	thirteen white and twelve blue	12
B2	thirteen white and twelve red	1
B3	thirteen blue and twelve white	1
B4	thirteen red and twleve blue	1
B5	green and white	5
B6	pink and white	2
B7	thirteen white and twelve blue plus blue border	10
B8	thirteen blue and twelve white plus blue border	4
B9	thirteen white and twelve red plus blue border	1

B10	thirteen blue and twelve white plus red border	4
B11	thirteen white and twelve blue plus red border	1
B12	thirteen white and twelve red plus red border	2
B13	twelve blue and nine white and four red plus red border	2
B14	twelve blue and nine white and four red plus blue border	1

Type C - ground with floweret or rosette (figure 13)

C1	blue ground plus four white hearts	6
C2	white ground plus four blue hearts	3
C3	white ground plus five blue petals	1
C4	blue ground plus six white petals	1
C5	white ground plus six blue petals plus red circle	3
C6	blue ground plus six white petals plus blue circle and white spot	1
C7	black ground plus six yellow petals plus red circle and white spot	1
C8	blue ground plus seven white petals plus white spot	1
C9	white ground plus seven blue petals plus blue circle	2
C10	yellow ground plus seven? red petals plus red circle	2
C11	blue ground plus eight white petals plus white spot	15
C12	blue ground plus eight white petals plus white circle plus red spot	1
C13	blue ground plus eight white petals plus red circle plus white spot	1
C14	blue ground plus eight dark blue petals plus white circle plus red spot	1
C15	white ground plus eight blue petals plus red circle	3
C16	red ground plus eight blue petals plus blue spot	1
C17	white ground plus eight blue petals plus red spot	1
C18	white ground plus eight blue petals plus blue circle	2
C19	white ground plus eight blue petals	1

Type D - miscellaneous patterns (figure 13)

D1	blue ground with one white whirl to right	5
D2	white ground with black fern (six pairs of arms)	2
D3	white ground with black fern (five pairs of arms)	1
D4	white ground with blue fern (six pairs of arms)	2
D5	white ground with blue fern (five pairs of arms)	2
D6	black ground with white cross	1
D7	alternate (?5) blue and white bars	1
D8	alternate white (3) and red (2) bars	2
D9	alternate green (3) and red (2) bars	1
D10	alternate yellow (2) and black bars	1
D11	alternate triangles of yellow and black	1
D12	sixteen squares blue and white plus darker blue border	1
D13	blue ground with sixteen white xx alternately	1
D14	white ring plus blue ring plus white crown (twelve points) plus red ring and yellow centre spot	4
D15	red spot with ten curved triangles alternately red and yellow fanning from it	1
D16	black ground with red square in centre and four white hearts	2
D17	white centre spot plus red circle plus six petals alternately blue and white plus red circle	1
D18	circle of alternate red and yellow rays (8 or 16?)	1
D19	four squares of four blue (2) and white (2) triangles with apices joined	4
D20	ditto but green and blue triangles	2

GLOSSARY

Base	the metal base to which the enamel is applied.
Bed	a layer of enamel of the one colour.
Cell	the recess gouged out (or cast) to take the enamel.
Champleve	gouged out - denotes that type of enamelwork where the cells are below the surface of the metal (gouged or cast); the term is avoided here as much as possible; all Romano-British enamels are champleve enamels.
Chequered	refers to millefiori glass/enamel where the pattern is made up of tiny alternate coloured squares.
Cloisonne	denotes enamelwork where the cells are above or on top of the metal surface; these are created by soldering thin strips of metal on to the surface.
Cloisson	the cell wall in cloisonne enamelwork but often wrongly applied to champleve enamelwork; the term is not used here if possible.
Complex (enamels)	where two or more different coloured enamels are applied to one cell without metal divides.
Decayed	where the original colour of an enamel is no longer discernible through internal or external factors.
Enamel	the actual vitreous substance whether raw lumps, powdered or in the form of slices; or slices of millefiori glass treated in this way.
Enamelwork	the end product consisting of enamel and metal.
Field	the predominant colour usually in the larger, outer cell on simple enamelwork and on complex enamelwork the bed or ground into which are laid spots of another colour.
Flux	an enamel used as an adhesive for, in particular, slices of millefiori glass.

Frit — generally means raw enamel especially powdered, sometimes used to denote raw glass often after the first melt when the colouring material has not yet been added to the basic ingredients.

'Blue Frit' — balls of a blue chalky substance, probably colouring matter but possibly unconnected with enamels.

Ground — the main layer or bed of enamel into which spots of another colour has been laid in the complex technique.

Inlay — general term used to describe the champleve or simple enamels of the Iron Age.

Millefiori (glass) — glass in the form of a rod with minute patterns of squares or flowerets etc. running right the way through it in the manner of sea-side rock.

Millefiori (enamel) — where slices of millefiori glass are used to decorate small objects and applied either directly by heat or floated on a bed of enamel.

Mosaic — a term to be avoided since it has been applied equally to what I term complex and millefiori enamels.

Paste — mixture of a powdered enamel of one colour and water, placed cold in the cell for firing as a simple enamel or as the ground for solid spots of a second colour in complex enamels.

Septum — the metal division between two cells in champleve enamelwork.

Setting — small circular cell found on Iron Age enamelwork.

Simple (enamels) — where only one colour of enamel is applied to the cell.

Uplay — general term used to describe the red studs on Iron Age enamelwork.

Weathered — see 'decayed'; can also mean where part or all of the enamel is missing or lost.

BIBLIOGRAPHY

Addyman, P.V. & Leigh, D. (1973) — "The Anglo-Saxon Village at Chalton, Hampshire: Second Interim Report", *Medieval Archaeology*, 17, 1973, pp.1-25

Agnew, R.V. (1881) — Donation by, of circular ornament from Dowalton Loch, *Proc. Society Antiquaries Scotland*, XV, 1880-1, pp.154-5

Akerman, J.W. (1860) — "Report on Researches in an Anglo-Saxon Cemetery at Long Wittenham, Berkshire, in 1859", *Archaeologia*, 38, 1860, pp.327-352

Alcock, L. (1963) — *Dinas Powys*, Cardiff, 1963

Alcock, L. (1971) — *Arthur's Britain*, London, 1971

Allen, J.R. (1904) — *Celtic Art In Pagan And Christian Times*, London, 1904

Allen, J.R. (1905) — "Find of Late-Celtic Bronze Objects at Seven Sisters, Near Neath, Glamorganshire", *Archaeologia Cambrensis*, 6th ser., v, 1905, pp.127-146

Amand, M. (1972) — *Vingt-cinq annees de fouilles archeologiques en Belgique*, Brussels, 1972

Anderson, J. (1881) — *Scotland In Early Christian Times*, Edinburgh, 1881

Anderson, J. (1884) — "Notice of An Enamelled Cup Or Patera Of Bronze Found In Linlithgowshire, Recently Purchased For The Museum", *Proc. Society Antiquaries Scotland*, XIX, 1884-5, pp.45-50

Anderson, J. (1900) — "Notice Of The Pottery, Bronze, And Other Objects Found At Camelon", *Proc. Society Antiquaries Scotland*, XXXV, 1900-01, pp.380-417

Anderson, J. (1905) — Notes on a Romano-British Hoard of Bronze Vessels and Personal Ornaments found in a Moss on Lamberton Moor, Berwickshire, *Proc. Society Antiquaries Scotland*, XXXIX, 1904-5, pp.367-378

Antiquaries Journal (1938) — "An Early British Enamel", unsigned note on, *Antiquaries Journal*, XVIII, 1938, pp.75-6

Apted, M.R. *et al.* eds. (1977) — *Ancient Monuments And Their Interpretation*, London, 1977

Archaeologia Cambrensis (1899) — "Archaeological Notes and Queries", *Archaeologia Cambrensis*, 5th ser., xvi, 1899, pp.259-267

Ashby, T. *et al.* (1904) — "Excavations at Caerwent, Monmouthshire, on the site of the Romano-British city of Venta Silurum, in the years 1901-3", *Archaeologia*, 59, 1904, pp.87-124

Ashby, T. *et al.* (1910) — Excavations at Caerwent, Monmouthshire, on the site of the Romano-British city of Venta Silurum, in the years 1909 and 1910", *Archaeologia*, 62, 1910, pp.405-448

Atkins, F.B. (1971) — "Egyptian Blue", pp.56-58 of Brodribb *et al.* (1971)

Atkinson, D. (1916) — *The Romano-British Site On Lowbury Hill In Berkshire*, Reading, 1916

Atkinson, D. (1942) — *Report on Excavations at Wroxeter 1923-7*, Oxford, 1942

Balch, H.E. (1929) — *Mendip - The Great Cave of Wookey Hole*, Wells, 1929

Ball, V. & Stokes, M. (1892) — "On a Block of Red Glass Enamel, said to have been found at Tara Hill. With Observations on the Use of Red Enamel in Ireland", *Trans. Royal Irish Academy*, XXX, 1892, pp.277-294

Barker, J. & Megaw, J.V.S. (1963) — "A Decorated Iron Age Bridle-bit in the London Museum: its place in Art and Archaeology", *Proc. Prehistoric Society*, XXIX, 1963, pp.206-213

Barnett, C. (1969) — *The Roman Caerwent Collection*, Newport, Museum and Art Gallery, 1969 (4th edition)

Bate, C.S. (1866) — "On The Discovery Of A Romano-British Cemetery Near Plymouth", *Archaeologia*, 40, 1866, pp.500-510

Bates, K.F. (1951) — *Enamelling: Principles and Practice*, New York, 1951

Bateson, J.D. (1973) — "Roman Material From Ireland: A Re-Consideration", *Proc. Royal Irish Academy*, 73, 1973, pp.21-97

Bateson, J.D. & Hedges, R. (1975) — "The Scientific Analyses of a Group of Roman-Age Enamelled Brooches",

Bateson & Hedges cont — Archaeometry, 17, 1975, pp.177-190

Bayley, J. (1978) — "The Roman Crucibles from Chapel Street", pp.254-5 of Down (1978)

Bayley, J. et al. (1980) — "The analysis of Roman Brooches from Richborough Fort, Kent", pp.239-247 of Slater & Tait eds. (1980)

Bedwin, O. (1980) — "Excavations at Chanctonbury Ring, Wiston, West Sussex, 1977", Britannia, XI, 1980, pp.173-222

Behrens, G. (1954) — "Zur Typologie Und Technik Der Provinzialromischen Fibeln", Jahrbuch Des Romisch-Germanischen Zentralmuseums Mainz, 1, 1951, pp.230-236

Bequet, A. (1902) — "La Bijouterie Chez Les Belges", Ann. Soc. Arch. de Namur, xxiv, 1902, pp.237-276

Biek, L. (1977) — "Enamel Residues and Similar Materials" p.70 of Butcher (1977)

Birley, E. (1931) — "Hadrian's Wall", Trans. Cumberland And Westmorland Antiquarian And Archaeological Society, new ser. 31, 1931, p.62

Birley, E. (1958) — "The Roman Fort At Brough-under-Stainmore", Trans. Cumberland And Westmorland Antiquarian And Archaeological Society, new ser. 58, 1958, pp.31-56

Boon, G. (1974) — Silchester The Roman Town Of Calleva, London, 1974

Boon, G. & Lewis, J.M. eds. (1976) — Welsh Antiquity, Cardiff, 1976

Boon, G. & Savory, H.N. (1975) — "A Silver Trumpet-Brooch with Relief Decoration, Parcel-gilt, from Carmarthen, and a Note on the Development of the Type", Antiquaries Journal, LV, 1975, pp.41-61

Borrajo, E.M. (1908) — Catalogue of the Collection of London Antiquities in the Guildhall Museum, London, 1908

Brailsford, J.W. (1962) — Hod Hill Vol. I Antiquities From Hod Hill In The Durden Collection, London, British Museum, 1962

Brailsford, J.W. (1964) — Guide To The Antiquities Of Roman Britain, London, British Museum, 1964

Brailsford, J.W. (1975) "The Polden Hill Hoard, Somerset", _Proc. Prehistoric Society_, XLI, 1975, pp.222-234

Branigan,K. (1977) _Gatscombe_, B.A.R. 44, Oxford, 1977

Brassington, M. (1980) "Derby Racecourse Kiln Excavations 1972-3", _Antiquaries Journal_, LX, 1980, pp.8-47

Breeze, D.J. (1974) "The Roman Fortlet at Barburgh Mill, Dumfriesshire", _Britannia_, V, 1974, pp.130-162

Brill, R.H. & Moll, S. (1963) "The Electron-Beam Micro-analysis Of Ancient Glass", pp.145-151 of _Recent Advances in Conservation_ ed. by G. Thompson, London, 1963

British Museum (1923) _Guide To Anglo-Saxon Antiquities_, London, British Museum, 1923

Brodribb, A. _et al_. (1968) _Excavations At Shakenoak I 1968_, Oxford, 1968

Brodribb, A. _et al_. (1971) _Excavations At Shakenoak II 1971_, Oxford, 1971

Brodribb, A. _et al_. (1972) _Excavations At Shakenoak III 1972_, Oxford, 1972

Brodribb, A. _et al_. (1973) _Excavations At Shakenoak IV_, Oxford, 1973

Brodribb, A. _et al_. (1978) _Excavations At Shakenoak V_, Oxford, 1978

Brown, D. (1976) "Archaeological Evidence for the Anglo-Saxon Period", pp.19-45 of McWhirr (1976)

Bruce, J. Collingwood (1880) "The Excavations at South Shields, Durham", _Archaeologia_, 46, 1880, pp.163-170

Bruce, J. Collingwood (1880) _Catalogue Of Antiquities At Alnwick Castle_, Newcastle-Upon-Tyne, 1880

Bruce, J. Collingwood (1885) "On the Recent Discoveries in the Roman Camp on The Lawe, South Shields", _Archaeologia Aeliana_, nex ser. x, 1885, pp.223-310

Bruce-Mitford, R. (1966) _The Sutton Hoo Ship Burial_, London, British Museum, 1966

Bruce-Mitford, R. (1974) _Aspects of Anglo-Saxon Archaeology_, London, 1974

Bruce-Mitford, R. (1975) _The Sutton Hoo Ship-Burial Vol. 1_, London, British Museum, 1975

Bruce-Mitford, R. (1978) — The Sutton Hoo Ship-Burial Vol. 2, London, British Museum, 1978

Bruce-Mitford, R. (1979) — The Sutton Hoo Ship-Burial A Handbook, London, British Museum, 1979

Budge, E.A. Wallis (1907) — An Account Of The Roman Antiquities Preserved In The Museum At Chesters, Northumberland, London, 1907

Bulleid, A. & Gray, H. St. George (1911) — The Glastonbury Lake Village, vol.I, Taunton, 1911

Bulliot, J.G. (1899) — Fouilles Du Mont Beuvray (Ancienne Bibracte) De 1887 A 1895, 2 vols., Autun, 1899

Bulmer, W. (1938) — "Dragonesque Brooches and their Development", Antiquaries Journal, XVIII, 1938, pp.146-153

Burley, E. (1955) — "A Catalogue And Survey Of The Metal-Work From Traprain Law", Proc. Society Antiquaries Scotland, LXXXIX, 1955-6, pp.118-226

Bushe-Fox, J.P. (1913) — Excavations on the Site of the Roman Town at Wroxeter, Shropshire, in 1912, Report Research Committee Society Antiquaries London no. 1, London, 1913

Bushe-Fox, J.P. (1914) — Second Report on the Excavations on the Site of the Roman Town at Wroxeter, Shropshire in 1913, Report Research Committee Society Antiquaries London no.II, London, 1914

Bushe-Fox, J.P. (1915) — Excavations at Hengistbury Head, Hampshire in 1911-2, Report Research Comittee Society Antiquaries London no. III, London, 1915

Bushe-Fox, J.P. (1916) — Third Report on the Excavations on the Site of the Roman Town at Wroxeter, Shropshire, in 1914, Report Research Committee Society Antiquaries London no. IV, London, 1916

Bushe-Fox, J.P. (1926) — First Report on the Excavation of the Roman Fort at Richborough, Kent, Report Research Committee Society Antiquaries London no. VI, Oxford, 1926

Bushe-Fox, J.P. (1949) — Fourth Report on the Excavation of the Roman Fort at Richborough, Kent, Report Research Committee Society Antiquaries London no. XVI, Oxford, 1949

Butcher, S.A. (1976) — "Enamelling" chapter 3, pp.42-51 of Strong & Brown eds. (1976)

Butcher, S.A. (1977) "Enamels From Roman Britain", pp.41-70 of Apted *et al*. (1977)

Butcher, S.A. (1978) "The Brooches" pp.216-220 of Jackson & Ambrose (1978)

Butler, R.M. ed. (1971) *Soldier and Civilian in Roman Yorkshire*, Leicester, 1971

Caley, E.R. (1962) *Analyses Of Ancient Glass 1790-1957*, New York, 1962

Callander, J.G. (1933) "A Collection of Prehistoric Relics From the Stevenson Sands, Ayrshire, And Other Objects in the National Museum", *Proc. Society Antiquaries Scotland*, LXVII, 1932-3, pp.26-34

Champion, S. (1976) "Coral in Europe: Commerse and Celtic Ornament", pp.29-40 of Duval & Hawkes eds. (1976)

Charles, J.A. (1975) "Where is the Tin ?", *Antiquity*, XLIX, 1975, pp.19-24

Charleston, R.J. (1963) "Glass 'Cakes' As Raw Material And Articles Of Commerse", *Journal Of Glass Studies*, 5, 1963, pp.54-67

Charlesworth, D. (1968) Note on 'Objects from the Roman Forts at Hardknott, Cumberland, and Brougham, Westmorland', *Antiquaries Journal*, XLVIII, 1968, p.306

Childe, V.G. (1933) "Excavations at Castlelaw Fort, Midlothian", *Proc. Society Antiquaries Scotland*, LXVII, 1932-3, pp.362-388

Childe, V.G. (1935) *The Prehistory Of Scotland*, London, 1935

Clarke, R.R. (1939) "The Iron Age in Norfolk and Suffolk", *Archaeological Journal*, xcvi, 1939, pp.1-113

Clarke, R.R. (1951) "A Hoard of Metalwork of the Early Iron Age from Ringstead, Norfolk", *Proc. Prehistoric Society*, XVII, 1951, pp.214-225

Clifford, E.M. (1961) *Bagendon: A Belgic Oppidum*, Cambridge, 1961

Cocks, A.H. (1921) "A Romano-British Homestead in the Hambleden Valley, Bucks.", Archaeologia, 71, 1921, pp.148-198

Coles, J.M. & Simpson, D.D.A. eds. (1968) *Studies in Ancient Europe*, Leicester, 1968

Collingwood, R.G. (1930) "Romano-Celtic Art In Northumberland", Archaeologia, 80, 1930, pp.37-58

Collingwood, R.G. (1930) The Archaeology Of Roman Britain, London, 1930

Collingwood, R.G. (1931a) "Roman Objects from Stanix and Thatcham", Antiquaries Journal, XI, 1931, pp.37-45

Collingwood, R.G. (1931b) "Roman Objects from Stanix", Trans. Cumberland and Westmorland A. & A. Society, new ser. 31, 1931, pp.69-80

Collingwood, R.G. (1931c) "Objects from Brough-under-Stainmore in the Craven Museum, Skipton", Trans. Cumberland and Wesrmorland A. & A. Society, new ser. 31, 1931, pp.81-86

Collingwood, R.G. & Richmond, I. (1969) The Archaeology Of Roman Britain, London, 1969

Corcoran, J. (1952) "Tankards and Tankard Handles of the British Early Iron Age", Proc. Prehistoric Society, XVIII, 1952, pp.85-102

Corder, P. (1935) Excavations At The Roman Fort At Brough, East Yorkshire, Hull Museum Publications no. 185, Hull, 1935

Corder, P. & Hawkes, C.F.C. (1940) "A Panel of Celtic Ornament from Elmswell, East Yorkshire", Antiquaries Journal, XX, 1940, pp.338-356

Cotton, A.R. (1933) "Saxon Discoveries at Fetcham", Archaeological Journal, xiii, 1933, pp.48-51

Cotton, M.A. (1947) "Excavations at Silchester 1938-9", Archaeologia, 92, 1947, pp.123-167

Cowen, J.D. & Richmond, I.A. (1935) "The Rudge Cup", Archaeologia Aeliana, 4th ser. XII, 1935, pp.310-342

Craddock, P. (1974a) "Roman Bronzes", Current Archaeology, iv, 1973-4, p.254

Craddock, P. (1974b) Letter re Roman Bronzes, Current Archaeology, iv, 1973-4, pp.349-350

Cramp, R. (1970) "Decorated Window-Glass and Millefiori From Monkwearmouth", Antiquaries Journal, L, 1970, pp.327-335

Cree, J.E. (1923) "Account of the Excavations on Traprain Law During the Summer of 1922", Proc. Society Antiquaries Scotland, LVII, 1922-3, pp.180-226

Cree, J.E. & Curle, A.O. (1922) — "Account of the Excavations on Traprain Law During the Summer of 1921", Proc. Society Antiquaries Scotland, LVI, 1921-2, pp.189-259

Cunliffe, B. (1964) — Winchester Excavations 1949-1960, vol. 1, Winchester, 1964

Cunliffe, B. (1968) — Fifth Report on the Excavations of the Roman Fort at Richborough, Kent, Report Research Committee Society Antiquaries London no. XXIII, Oxford, 1968

Cunliffe, B. (1969) — Roman Bath, Report Research Committee Society Antiquaries London no. XXIV, Oxford, 1969

Cunliffe, B. (1971) — Excavations At Fishbourne 1961-9 Vol. II The Finds, Report Research Committee Society Antiquaries London no. XXVIII, Leeds, 1971

Cunliffe, B. (1978) — Hengistbury Head, London, 1978

Curle, A.O. (1914) — "Report on the Excavation,in September 1913, of a Vitrified Fort at Rockcliffe, Dalbeattie, known as the Mote of Mark", Proc. Society Antiquaries Scotland, XLVIII, 1913-14, pp.125-168

Curle, A.O. (1915) — "Account of Excavations on Traprain Law in the Parish of Prestonkirk, County of Haddington, in 1914", Proc. Society Antiquaries Scotland, XLIX, 1914-15, pp.139-202

Curle, A.O. & Cree, J.E. (1916) — "Account of the Excavations on Traprain Law in the Parish of Prestonkirk, County of Haddington, in 1915", Proc. Society Antiquaries Scotland, L, 1915-16, pp.64-144

Curle, A.O. & Cree, J.E. (1921) — "Account of the Excavations on Traprain Law During the Summer of 1920", Proc. Society Antiquaries Scotland, LV, 1920-21, pp.153-206

Curle, C.L. (1974) — "An Engraved Lead Disc From The Brough Of Birsay, Orkney", Proc. Society Antiquaries Scotland, CV, 1972-4, pp.301-307

Curle, J. (1911) — A Roman Frontier Post And Its People (The Fort of Newstead in the Parish of Melrose), Glasgow, 1911

Curle, J. (1932) — "An Inventory of Objects of Roman and Provincial Roman Origin Found on Sites in Scotland Not Definitely Associated with

Curle, J. (1932) cont — Roman Constructions", Proc. Society Antiquaries Scotland, LXVI, 1931-2, pp.277-397

Current Archaeology (1974) — Correspondence re Roman Bronzes, Current Archaeology, iv, 1973-4, pp.349-350

Dalton, O.M. (1912) — Catalogue Of The McClean Bequest, Fitzwilliam Museum, Cambridge, 1912

Davies, E. (1949) — The Prehistoric And Roman Remains Of Flintshire, Cardiff, 1949

Davis, J.L. & Spratling, M. (1976) — "The Seven Sisters Hoard: a centenary study", pp.121-147 of Boon & Lewis eds.(1976)

Dawkins, W. Boyd (1874) — Cave Hunting, London, 1874

Dechelette, J. (1927) — Manuel D'Archeologie vol. iv Second Age Du Fer Du Epoque De La Tene, Paris, 1927

Dickinson, T.M. (1979) — "On the Origin and Chronology of the Early Anglo-Saxon Disc Brooch", pp.39-80 of Hawkes et al. eds. (1979)

Down, A. (1978) — Chichester Excavations III, Chichester, 1978

Down, A. & Rule, M. (1971) — Chichester Excavations I, Chichester, 1971

Dudley, D. (1967) — "Excavations on Nor Nour in the Isles of Scilly, 1962-6", Archaeological Journal, cxxiv, 1967, pp.1-64

Dudley, D. & Jope, E.M. (1965) — "An Iron Age Cist-Burial with Two Brooches from Trevone, North Cornwall", Cornish Archaeology, 4, 1965, pp.18-23

Durham, D. (1864) — Donation by, of collection of silver ornaments from Norries' Law, Proc. Society Antiquaries Scotland, VI, 1864-6, pp.7-11

Duval, P-M. & Hawkes, C. eds. (1976) — Celtic Art In Ancient Europe Five Protohistoric Centuries, London, 1976

Edinburgh (1892) — Catalogue of the National Museum of Antiquities of Scotland, Edinburgh, 1892

Ellis, H. (1829) — "Observations upon an ancient Bracelet of bronze, found upon the sandhills near Altyre, on the coast of Murrayshire", Archaeologia, 22, 1829, pp.285-296

Exner, K. Von (1939) — "Die Provinzialromischen Emailfiblen der Rheinlande", Bericht Der Romisch-Germanischen Kommission, 29, 1939, pp.31-121

Feachem, R.W. (1951) "Dragonesque Fibulae", Antiquaries Journal, XXXI, 1951, pp.32-44

Feachem, R.W. (1968) "Dragonesque Fibulae", Antiquaries Journal, XLVIII, 1968, pp.101-102

Filip, J. (1960) Celtic Civilisation and its Heritage, Prague, 1960

Flower, J. (1880) "On the Process of Decay in Glass", Archaeologia, 46, 1880, pp.65-162

Forbes, R.J. (1957) Studies in Ancient Technology, "Glass", vol. v, pp.110-231, Leiden, 1957

Fowler, E. (1963) "Celtic Metalwork of the Fifth and Sixth Centuries A.D.", Archaeological Journal, cxx, 1963, pp.98-160

Fowler, E. (1968) "Hanging Bowls", pp.287-310 of Coles & Simpson eds. (1968)

Fowler, P.J. (1965) "A Roman Barrow at Knob's Crook, Woodlands, Dorset", Antiquaries Journal, XLV, 1965, pp.22-52

Fox, A. (1952) Roman Exeter, Manchester, 1952

Fox, A. & Pollard, S. (1973) "A Decorated Bronze Mirror From An Iron Age Settlement at Holcombe, Near Uplyme, Devon", Antiquaries Journal, LIII, 1973, pp.16-41

Fox, C. (1923) The Archaeology Of The Cambridge Region, Cambridge, 1923

Fox, C. (1930) "A La Tene Weight from Winchester", Antiquaries Journal, X, 1930, pp.251-253

Fox, C. (1946) A Find Of The Early Iron Age From Llyn Cerrig Bach, Anglesey, National Museum of Wales, Cardiff, 1946

Fox, C. (1948) "The Incised Ornament On The Celtic Mirror From Colchester, Essex", Antiquaries Journal, XXVIII, 1948, pp.123-137

Fox, C. (1958) Pattern and Purpose, National Museum of Wales, Cardiff, 1958

Fox, G.E. (1894) "Excavation on the site of the Roman City of Silchester, Hants., in 1894", Archaeologia, 56, 1894, pp.439-494

Fox, G.E. (1900) "Roman Suffolk", Archaeological Journal, lvii, 1900, pp.89-165

Franks, A.W. (1854) "The Additions to the Collections of National Antiquities in the British Museum", Archaeological Journal, xi, 1854, pp.23-32

Franks, A.W. (1870) "Note on Roman Enamelled Cup from Braughing and others of this type", Proc. Society Antiquaries London, 2nd ser. iv, 1867-70, pp.514-516

Frazer-Tytler, N. (1923) Exhibition by, of enamelled belt-plate and dress-fastener from Drumashie, Proc. Society Antiquaries Scotland, LVIII, 1923-4, pp.11-13

Frere, S.S. (1972) Verulamium Excavations Vol. I, Report Research Committee Society Antiquaries London no. XXVIII, Oxford, 1972

Frere, S.S. (1974) Britannia, London (Cardinal edition), 1974

Gage, J. (1836) "A Letter from John Gage, Esq. F.R.S., Director, to Hudson Gurney, Esq., F.R.S., Vice-President, communicating the recent discovery of Roman sepulchral relics in one of the greater Barrows at Bartlow, in the parish of Ashdon, in Essex", Archaeologia, 26, 1836, pp.300-317

Gardener, W. & Savory, H.N. (1964) Dinorben, Cardiff, 1964

Gillam, J.P. (1958) "Roman and Native, A.D. 122-197", chapter III, pp.60-90 of Richmond (1958)

Goodburn, R. (1979) "Roman Britain in 1978 1. Sites Explored", Britannia, X, 1979, pp.268-338

Gray, H. St. George & Bulleid, A. (1953) The Meare Lake Village, Vol. II, Taunton, 1953

Green, B. (1972) "Three Iron Age Terrets From Norfolk", Antiquaries Journal, LII, 1972, p.346

Green, C. (1949) "The Birdlip Early Iron Age Burials: a Review", Proc. Prehistoric Society, XV, 1949, pp.188-190

Green, M.J. (1975) "Romano-British Non-Ceramic Model Objects in South-east Britain", Archaeological Journal, cxxxii, 1975, pp.54-70

Green, M.J. (1976) A Corpus Of Religious Material From The Civilian Areas Of Roman Britain, B.A.R. 24, Oxford, 1976

Green, M.J. (1978) A Corpus Of Small Cult-Objects From The

Green, M.J. (1978) cont — *Military Areas Of Roman Britain*, B.A.R. 52, Oxford, 1978

Greenfield, E. (1963) — "The Romano-British Shrines At Brigstock, Northants.", *Antiquaries Journal*, XLIII, 1963, pp.228-263

Grigson, W.H. (1849) — Presentation by, of the Saham Toney hoard, *Norfolk Archaeology*, ii, 1849, pp.398-400

Grimes, W.F. (1930) — *Holt, Denbighshire*, London, 1930 (vol. xli of *Y Cymmodor*)

Grimes, W.F. (1951) — *Aspects of Archaeology in Britain and Beyond*, London, 1951

Guido, M. (1978) — *The Glass Beads of the Prehistoric and Roman Periods in Britain and Ireland*, Report Research Committee Society Antiquaries London no. XXXV, London, 1978

Guildhall Museum (1960) — *Small Finds From The Walbrook 1954-1955*, London, 1960

Gunther, R.T. (1912) — "A Mural Glass Mosaic from the Imperial Villa, near Naples", *Archaeologia*, 63, 1912, pp.99-105

Harden, D.B. (1956) — "Glass and Glazes", chapter 9, pp.311-346 of Singer *et al*. (1956)

Harden, D.B. (1968) — "Ancient Glass, I: Pre-Roman", *Archaeological Journal*, cxxv, 1968, pp.46-72

Harden, D.B. (1969) — "Ancient Glass, II: Roman", *Archaeological Journal*, cxxvi, 1969, pp.44-77

Harden, D.B. (1971) — "Ancient Glass, III: Post-Roman", *Archaeological Journal*, cxxviii, 1971, pp.78-117

Harding, D.W. (1972) — *The Iron Age Of The Upper Thames Basin*, Oxford, 1972

Harding, D.W. (1974) — *The Iron Age In Lowland Britain*, London, 1974

Harding, J.R. (1960) — "Two Frankish Beads from the Coast of Tanganyika", *Medieval Archaeology*, 4, 1960, pp.126-127

Harford, C.J. (1803) — "An Account of Antiquities found in Somersetshire", *Archaeologia*, 14, 1803, pp.90-93

Harrod, H. (1855) — "On Horse-trappings found at Westhall", Archaeologia, 36, 1855, pp.454-456

Hartley, B.R. (1954) — "Bronze-worker's Hearth", Journal Chester and North Wales Architectural, Archaeological and Historical Society, 41, 1954, pp.1-14

Haseloff, G. (1958) — "Fragments of a Hanging-Bowl from Bekesbourne, Kent, and Some Ornamental Problems", Medieval Archaeology, 2, 1958, pp.72-103

Hawkes, C. (1927) — "Excavations at Alchester, 1926", Antiquaries Journal, VII, 1927, pp.147-184

Hawkes, C.F.C. (1935) — "A La Tene bronze handle from Welwyn", Antiquaries Journal, XV, 1935, pp.351-354

Hawkes, C.F.C. (1947) — "Britons, Romans and Saxons Round Salisbury And In Cranborne Chase", Archaeological Journal, civ, 1947, pp.27-81

Hawkes, C.F.C. & Hull, M.R. (1947) — Camulodunum, Report Research Committee Society Antiquaries London no.XIV, Oxford, 1947

Hawkes, C.F.C. (1951) — "Bronze-workers, Cauldrons, and Bucket-animals in Iron Age and Roman Britain", pp.192-199 of Grimes (1951)

Hawkes, C.F.C (1959) — "The ABC of the British Iron Age", Antiquity, 33, 1959, pp.170-182

Hawkes, S.C. et al. eds. (1979) — Anglo-Saxon Studies in Archaeology and History I, B.A.R. 72, Oxford, 1979

Hencken, H. (1951) — "Lagore Crannog: an Irish royal residence of the 7th-10th centuries A.D.", Proc. Royal Irish Academy, 53, c, 1950-51, pp.1-247

Henderson, J. (1980) — "Some new evidence for iron age glass-working in Britain", Antiquity, LIV, 1980, pp.60-61

Henry, F. (1933) — "Emailleurs D'Occident", Prehistoire, ii, 1933, pp.66-146

Henry, F. (1936) — "Hanging Bowls", Journal Royal Society Antiquaries Ireland, lxvi, 1936, pp.209-246

Henry, F. (1956) — "Irish Enamels of the Dark Ages and their Relation to the Cloisonne Technique", pp.71-90 of Dark Age Britain, ed. D.B. Harden, London, 1956

Heslop, R.B. & Robinson, P.L. (1963) — Inorganic Chemistry, New York, 1963

Heurgon, J. (1951) "The Amiens Patera", Journal Roman Studies, xli, 1951, pp.22-24

Hewitt, A.T.M. (1968) Roman Villa, West Park, Rockbourne, Nr. Fordingbridge, Hants., Illustrated Report, privately, 1968

Hildyard, E.J.W. (1954) "The Rudston Fibulae", Antiquaries Journal, XXXIV, 1954, pp.73-75

Hills, G.M. (1871) Description by, of three ancient brooches, Journal British Archaeological Association, 27, 1871, pp.258-259

Hinton, D.A. (1974) Catalogue Of The Anglo-Saxon Ornamental Metalwork 700-1100 In The Department Of Antiquities Ashmolean Museum, Oxford, 1974

Hobley, B. (1969) "A Neronian-Vespasianic Military Site At 'The Lunt', Baginton, Warwickshire", Birmingham Archaeological Society Trans. & Proc., 83, 1966-7, pp.65-129

Hobley, B. (1973) "Excavations At 'The Lunt' Roman Military Site, Baginton, Warwickshire, 1968-71. Second Interim Report", Birmingham And Warwickshire Archaeological Society Trans., 85, 1971-3, pp.7-92

Hodges, H. (1964) Artefacts, London, 1964

Hogg, A.H.A. (1951) "The Votadini", pp.200-220 of Grimes (1951)

Hughes, M.J. (1972) "A technical study of opaque red glass of the Iron Age in Britain", Proc. Prehistoric Society, XXXVIII, 1972, pp.98-107

Hull, M.R. (1958) Roman Colchester, Report Research Committee Society Antiquaries London no. XX, Oxford, 1958

Hull, M.R. (1964) "Brooches", pp.88-90 of Cunliffe (1964)

Hull, M.R. (1967) "The Nor Nour Brooches", pp.28-64 of Dudley (1967)

Hunter, A.G. & Kirk, J.R. (1952) "Excavations at Campsfield, Kidlington, Oxon., 1949", Oxoniensia, xvii, 1952, pp.36-62

Jackson, D.A. & Ambrose, T.M. (1978) "Excavations at Wakerly, Northants, 1972-75", Britannia, IX, 1978, pp.115-242

Jacobsthal, P. (1944) Early Celtic Art, Oxford, 1944

Jenkins, F. (1951) Roman Canterbury No. 5. An Account of the Excavations in Burgate in 1946-8, London, 1951

Jessup, R. (1950) — *Anglo-Saxon Jewellery*, London, 1950

Jessup, R. (1970) — *South-East England*, London, 1970

Johns, C. (1974) — "A Roman silver pin from Oldcroft, Gloucestershire", *Antiquaries Journal*, LIV, 1974, pp.295-297

Jones, G.D.B. (1974) — *Roman Manchester*, Manchester, 1974

Jones, M.U. *et al*. (1976) — "Celtic coin-moulds from Old Sleaford, Lincolnshire", *Antiquaries Journal*, LVI, 1976, pp.238-241

Jope, E.M. (1951) — "A Bronze Butt Or Ferrule From The River Bann", *Belfast Municipal Museum and Art Gallery Bulletin*, i, 1951, pp.57-65

Jope, E.M. (1955) — "Chariotry and Paired-Draught in Ireland during the Early Iron Age: The Evidence of some Horse-Bridle-Bits", *Ulster Journal Archaeology*, 3rd ser. 18, 1955, pp.37-44

Jope, E.M. (1971) — "The Witham Shield", pp.61-69 of Sieveking (1971)

Jope, E.M. & Wilson, B.C.S. (1957) — "A Burial Group of the First Century A.D. from 'Loughey', near Donaghadee, Co. Down", *Ulster Journal Archaeology*, 3rd ser. 20, 1957, pp.73-94

Kemble, J.M. (1863) — *Horae Ferales*, London, 1863

Kendrick, T.D. (1932) — "British Hanging-Bowls", *Antiquity*, vi, 1932, pp.161-184

Kendrick, T.D. (1941) — "A Late Saxon Hanging-bowl", *Antiquaries Journal*, XXI, 1941, pp.161-162

Kendrick, T.D. (1941a) — "The Scunthorpe Bowl", *Antiquaries Journal*, XXI, 1941, pp.236-238

Kenyon, K. (1934) — "The Roman Theatre at Verulamium, St. Albans", *Archaeologia*, 84, 1934, pp.213-261

Kenyon, K.M. (1938) — "Excavations at Viroconium, 1936-7" *Archaeologia*, 88, 1938, pp.213-288

Kenyon, K.M. (1948) — *Excavations at the Jewry Wall Site, Leicester*, Report Research Committee Society Antiquaries London no. XV, Oxford, 1948

Kilbride-Jones, H.E. (1937) — "The Evolution Of Penannular Brooches With Zoomorphic Terminals In Great Britain

Kilbride-Jones, H.E. (1937) cont. — And Ireland", Proc. Royal Irish Academy, 43, 1935-7, pp.379-455

Kilbride-Jones, H.E. (1980a) — Zoomorphic Penannular Brooches, Report Research Committee Society Antiquaries London no. XXXIX, London, 1980

Kilbride-Jones, H.E. (1980b) — Celtic Craftsmanship In Bronze, London, 1980

King, A.(1971) — "Romano-British Metalwork From The Settle District Of West Yorkshire", Yorkshire Archaeological Journal, xlii, 1971, pp.410-417

Kirk, J.R. (1949) — "Bronzes from Woodeaton, Oxon.,", Oxoniensia, xiv, 1949, pp.1-45

Kramer, W. (1960) — "The Oppidum of Manching", Antiquity, xxxiv, 1960, pp.1-20

Laing, L. (1975) — "The Mote of Mark and the origins of Celtic interlace", Antiquity, xlix, 1975, pp.98-108

Laver, P.G. (1926) — "The Excavation of a Tumulus at Lexden, Colchester", Archaeologia, 78, 1927, pp.241-254

Leeds, E.T. (1933) — Celtic Ornament, Oxford, 1933

Leeds, E.T. (1949) — Early Anglo-Saxon Great Square-Headed Brooches, Oxford, 1949

Lerat, L. (1956) — Catalogue Des Collections Archeologiques De Besancon II Les Fibules Gallo-Romaines, Besancon, 1956

Lethbridge, T.C. (1952) — "Excavations at Kilpedher, South Uist, and the problem of Brochs and Wheel-houses", Proc. Prehistoric Society, XVIII, 1952, pp.176-193

Lethbridge, T.C. (1952) — "Bronze Bowl of the Dark Ages from Hildersham, Cambs.", Proc. Cambridge Antiquarian Society, xlv, 1952, pp.44-47

Loe, Le Baron de (1937) — Belgique Ancienne: Catalogue Descriptif Et Raisonne, III La Periode Romaine, Brussels, 1937

London Museum (1962) — Roman London, London, 1962 (compiled by H.W. Young)

Longley, D. (1975) — The Anglo-Saxon Connection, B.A.R. 22, Oxford, 1975

Longley, D. (1980)	The Mote Of Mark: a Dark Age Hillfort in south-west Scotland (1979 excavations: interim report), St. Andrews, 1980 (hand-out)
Longworth, I. (1965)	"A Massive Cist with Multiple Burials of Iron Age Date at Lochend,Dunbar", Proc. Society Antiquaries Scotland, XCVIII, 1964-6, pp. 173-183
Lowry, P.R. *et al*. (1971)	"Scriber, Graver, Scorer, Tracer; notes on Experiments in Bronze-working Technique", Proc. Prehistoric Society, XXXVIII, 1971, pp.167-182
Lucas, A. (1948)	Ancient Egyptian Materials and Industries, London, 1948
MacDonald, G. & Curle, J.O. (1929)	"The Roman Fort At Mumrills Near Falkirk", Proc. Society Antiquaries Scotland, LXIII, 1928-9, pp.396-575
MacGregor, G. (1962)	"The Early Iron Age Metalwork Hoard From Stanwick, N.R. Yorks.", Proc. Prehistoric Society, XXVIII, 1962, pp.17-57
McWhirr, A. ed. (1976)	Studies in the Archaeology and History of Cirencester, B.A.R. 30, Oxford, 1976
Marsden, P. (1980)	Roman London, London, 1980
Maryon, H. (1954)	Metalwork and Enamelling, London, 1954
Maryon, H. (1956)	"Fine Metal-Work", chapter 13 of Singer *et al*. (1956)
Maxwell, H.E. (1886)	"Notice of an Enamelled Bronze Harness Ornament from Auchendolly, Stewartry of Kilcudbrightshire", Proc. Society Antiquaries Scotland, XX, 1885-6, pp.396.398
May, T. (1904)	Warrington's Roman Remains, Warrington, 1904
May, J. (1971)	"An Iron Age Spout From Kirmington, Lincolnshire", Antiquaries Journal, LI, 1971, pp.252-259
Megaw, J.V.S. (1963)	"A British Bronze Bowl Of The Belgic Iron Age From Poland", Antiquaries Journal, XLIII, 1963, pp.27-37
Megaw, J.V.S. (1970)	Art Of The European Iron Age, Bath, 1970

Megaw, J.V.S. (1971) "A Group Of Later Iron Age Collars Or Neck-Rings From Western Britain", pp.145-156 of Sieveking (1971)

Megaw, J.V.S. (1971) "The Felmersham Fish-Head Spout", Antiquaries Journal, LI, 1971, pp.299-300

Merriefield, R. (1962) "Coins From The Bed Of The Walbrook, And Their Significance", Antiquaries Journal, XLII, 1962, pp.38-52

Merriefield, R. (1965) The Roman City Of London, London, 1965

Michon, E. (1925) "Anneau emaille trouve a Eauze (Gers)", Antiquaries Journal, V, 1925, pp.145-147

Miket, R. (1974) "Metalwork" pp.159-162 of Breeze (1974)

Moore, C.N. (1973) "Two Examples of Late Celtic and Early Roman Metalwork from South Lincolnshire", Britannia, IV, 1973, pp.153-159

Moore, C.N. (1978) "An enamelled Skillet-handle from Brough-on-Fosse", Britannia, IX, 1978, pp.319-326

Morrison, J. (1873) Donation by, of pin with enamelled head from Urquhart, Proc. Society Antiquaries Scotland, X, 1872-4, p.357

Mottram, A.S. (1968) "An Enamelled Bronze Armlet From Southacre", Norfolk Archaeology, 34, 1968, pp.259-262

Munro, R. (1899) "Notes on a Crannog at Hyndford, near Lanark, recently discovered and excavated by Andrew Smith", Proc. Society Antiquaries Scotland, XXXIII, 1898-9, pp.373-387

Musty, J. & Butcher, S. (1976) "A reassessment of Roman 'bronzes' from Richborough Roman Fort, Kent", Antiquaries Journal, LXVI, 1976, pp.241-242

Nash-Williams, V.E. (1930) "Further Excavations at Caerwent, Monmouthshire, 1923-5", Archaeologia, 80, 1930, pp.229-288

Neal, D.S. (1974) The Excavation of the Roman Villa in Gadebridge Park, Hemel Hempstead, Report Research Committee Society Antiquaries London no. XXXI, London, 1974

Newton, R.G. (1971) "A Preliminary Examination Of A Suggestion That Pieces Of Strongly Coloured Glass Were Articles Of Trade In Iron Age Britain", Archaeometry, 13, 1971, pp.11-16

Olivier, A.C.H. (1979) "Brooches" pp.65-69 (Ravenglass) and 206-212 (Watercock) of Potter (1979)

O'Riordain, S.P. (1942) "The Excavation Of A Large Earthen Ring-Fort At Garranes, Co. Cork", *Proc. Royal Irish Academy*, 47, 1941, pp.77-150

Painter, K. & Sax, M. (1970) "The British Museum collection of Roman head-stud brooches", *British Museum Quarterly*, xxxiv, 1970, pp.153-174

Peers, C. & Radford, C.A.R. (1943) "The Saxon Monastery of Whitby", *Archaeologia*, 89, 1943, pp.27-88

Piggott, S. (1950) "Excavations in the Broch and Hill-Fort Of Torwoodlee, Selkirkshire, 1950", *Proc. Society Antiquaries Scotland*, LXXXV, 1950-1, pp.92-137

Piggott, S. (1952) "Three Metalwork Hoards Of The Roman Period From Southern Scotland", *Proc. Society Antiquaries Scotland*, LXXXVI, 1952-3, pp.1-50

Piggott, S. (1970) *Early Celtic Art*, Edinburgh, 1970

Pitt Rivers, A. (1887) *Excavations In Cranborne Chase Vol. I*, privately, 1887

Pitt Rivers, A. (1888) *Excavations In Cranborne Chase Vol. II*, privately, 1888

Pitt Rivers, A. (1892) *Excavations In Bokerly and Wansdyke, Dorset and Wilts. 1888-1891 Vol. III*, privately, 1892

Pitt Rivers, A. (1898) *Excavations In Cranborne Chase Vol. IV*, privately, 1898

Pitts, L.F. (1979) *Roman Bronze Figurines from the Civitates of the Catuvellauni and Trinovantes*, B.A.R. 60, Oxford, 1979

Potter, T. (1977) "The Biglands Milefortlet and the Cumberland Coast Defenses", *Britannia*, VIII, 1977, pp.149-183

Potter, T. (1979) *Romans In North-West England*, Kendal, 1979

Price, J.E. (1873) *Roman Antiquities Recently Discovered On The Site Of The National Safe Deposit Company's Premises, Mansion House, London*, London, 1873

Preston, H. (1916) — "Roman British Remains At Saltersford, near Grantham", *Lincolnshire Notes and Queries*, xiv, 1916-7, pp.35-50

P.S.A.S. (1901) — Purchase of enamelled bronze pin from Pabbay, *Proc. Society Antiquaries Scotland*, XXXV, 1900-01, p.278

P.S.A.S. (1908) — Purchase of enamelled pin from Birsay, *Proc. Society Antiquaries Scotland*, XLIII, 1908-9, pp.8-9

P.S.A.S. (1930) — Purchase of enamelled hand pin, *Proc. Society Antiquaries Scotland*, LXV, 1930-31, p.13

P.S.A.S. (1947) — Purchase of enamelled silver pin from Freswick Links, *Proc. Society Antiquaries Scotland*, LXXXI, 1946-7, p.196

Rae, A. & V. (1974) — "The Roman Fort at Cramond, Edinburgh: Excavations 1954-1966", *Britannia*, V, 1974, pp.163-224

Read, C.H. (1910) — Exhibition of bridle-bit from Thames, now in B.M., *Proc. Society Antiquaries*, 2nd ser. xxiii, 1909-11, pp.156-159

Rhatz, P.A. & Greenfield, E. (1977) — *Excavations At Chew Valley Lake*, London, 1977

Richardson, K.M. (1944) — "Report on Excavations at Verulamium: Insula XVII, 1938", *Archaeologia*, 90, 1944, pp.81-126

Richardson, K.M. (1960) — "A Roman Brooch from the Outer Hebrides with Notes on Others of its Type", *Antiquaries Journal*, XL, 1960, pp.200-213

Richardson, K.M. (1969) — "A Brooch From The Pump Room Hotel", pp.64-65 of Cunliffe (1969)

Richmond, I.A. (1958) — *Roman And Native In North Britain*, Edinburgh, 1958

Richmond, I.A. (1968) — *Hod Hill Vol. 2 Excavations Carried Out Between 1951 and 1956*, London, 1968

Riha, E. (1979) — *Die Romischen Fibeln Aus Augst Und Kaiseraugst*, Augst, 1979

Robertson, A.S. (1970) — "Roman Finds From Non-Roman Sites In Scotland", *Britannia*, I, 1970, pp.198-226

Royal Commission (1962) — *Eburacum*, vol. I, London, Royal Commission Historic Monuments England, 1962

Saleh, S.A. et al. (1972) "Study of glass and glass-making processes at Wadi El-Natrum, Egypt, in the Roman period 30 B.C. to A.D. 359. Part I Fritting crucibles, their technical features and temperature employed", Studies In Conservation, 17, 1972, pp.143-172

Savory, H.N. (1966) "A Find Of Early Iron Age Metalwork From From The Lesser Garth, Pentyrch (Glam.)", Archaeologia Cambrensis, cxv, 1966, pp.27-44

Savory, H.N. (1976) Guide Catalogue of the Early Iron Age Collections, Cardiff, National Museum of Wales, 1976

Sayre, E.V. (1965) "Summary of the Brookhaven Program of Analysis of Ancient Glass", pp.145-154 of Application of Science in the Examination of Works of Art, Boston, 1965

Scott, J. (1922) Donation to National Museum of Antiquities by, of collection from Trentsmuir, St. Andrews, Fife., Proc. Society Antiquaries Scotland, LVII, 1922-3, p.109

Sellye, I. (1939) "Les Bronze Emailles de la Pannonie Romaine", Dessertationes Pannonicae, ser. 2, fasc. 8, 1939, pp.1-91

Shaw, K. (1972) Science for Craft Potters and Enamellers, Newton Abbot, 1972

Simpson, G. (1964) Britons and The Roman Army, London, 1964

Simpson, M. (1968) "Massive Armlets in the North British Iron Age", pp.233-254 of Coles & Simpson (1968)

Simpson, M. (1970) "Some Roman-Iron Age Finger Rings", Proc. Society Antiquaries Scotland, CII, 1964-70, pp.105-108

Simpson, M. (1972) "An Iron Age Tankard Fragment From Caerwent", Antiquaries Journal, LII, 1972, pp.330-331

Simpson, W.D. (1954) Dundarg Castle, Edinburgh, 1954

Singer, C. et al. (1958) A History of Technology, vol. III, Oxford, 1956

Slater, E.A. & Tait, J.O. (1980) Proceedings Of The 16th International Symposium On Archaeometry And Archaeological Prospection Edinburgh 1976, Edinburgh, 1980

Small, A. & Cottam, M.B. (1972) Craig Phadrig, Dundee, 1972

Small, A. *et al.* (1973) — *St. Ninian's Isle And Its Treasure*, Oxford, 1973

Smith, D.L. (1976) — "Survey of Badge Enamelling in Birmingham", *Medical Science Law*, 16, 1976, pp.33-36

Smith, H.E. (1866) — "Some Interesting Features Of A Neglected Roman Station, Brough-Under-Stainmore", *Trans. Historical Society Lancashire & Cheshire*, new ser. vi, 1865-6, pp.137-152

Smith, J.A. (1881) — "Notice Of A Massive Bronze 'Late-Celtic' Armlet And Two Small Objects Of Bronze (Horse-Harness) Found With A Roman Bronze Patella, At Stanhope, Peeblesshire, In 1876; With An Account Of Other Bronze Or Brass Armlets Found In Scotland", *Proc. Society Antiquaries Scotland*, XV, 1880-1, pp.316-363

Smith, J. (1919) — "Excavation Of The Forts Of Castlehill, Aitnock, and Coalhill, Ayrshire", *Proc. Society Antiquaries Scotland*, LIII, 1918-9, pp.123-134

Smith, R.A. (1908) — "Note on a Romano-British hoard of bronze objects found on Lamberton Moor, Berwickshire", *Proc. Society Antiquaries Scotland*, XXII, 1907-9, pp.51-62

Smith, R.A. (1908) — "Note and List of Hanging Bowls", *Proc. Society Antiquaries London*, 2nd ser. xxii, 1907-9, pp.66-86

Smith, R.A. (1909) — "A Hoard of Metal Found at Santon Downham, Suffolk", *Proc. Cambridge Antiquarian Society*, xiii, 1908-9, pp.146-163

Smith, R.A. (1925) — *Guide To Early Iron Age Antiquities*, London, British Museum, 1925

Smythe, J.A. (1937) — "Roman Objects Of Copper And Iron From The North Of England", *Proc. University Durham Philosophical Society*, ix, 1931-7, pp.382-405

Spitaels, P. (1970) — "La Villa Gallo-Romaine D'Anthee", *Helenium*, x, 1970, pp.209-241

Spratling, M. (1970) — "The Late Pre-Roman Iron Age Bronze Mirror From Old Warden", *Bedfordshire Archaeological Journal*, v, 1970, pp.5-16

Spratling, M. (1975) — "Fragments of a lorica segmentata in the hoard from Santon, Norfolk", *Britannia*, VI, 1975, pp.206-207

Spratling, M. (1979) "The Debris of Matal Working" chapter IX, pp.125-149 of Wainwright (1979)

Spratling, M. (1980) "Observations on the Analysis of the Composition of the Opaque Red Glass ('Red Enamel') of the British Pre-Roman Iron Age", pp.113-118 of Slater & Tait eds. (1980)

Spratling, M. (1980) "An Iron Age Bronze Foundry At Gussage All Saints, Dorset: Preliminary Assessment Of Technology", pp.268-292 of Slater & Tait eds. (1980)

Stead, I.M. (1965) The La Tene Cultures Of Eastern Yorkshire, York, 1965

Stead, I.M. (1967) "A La Tene III Burial at Welwyn Garden City", Archaeologia, 101, 1967, pp.1-62

Stead, I.M. (1971) "Yorkshire before the Romans:some recent discoveries", pp.1-43 of Butler (1971)

Stead, I.M. (1976) Excavations At Winterton Roman Villa, London, 1976

Stead, I.M. (1979) The Arras Culture, York, 1979

Strong, D. & Brown, D. eds. (1976) Roman Crafts, London, 1976

Stevenson, R.B.K. (1972) "Note on the mould from Craig Phadrig", pp.49-51 of Small & Cottam (1972)

Taylor, J.R. (1977) "The Origin And Use Of Cobalt Compounds As Blue Pigments", Science And Archaeology, 19, 1977, pp.3-15

Taylor, M.V. (1917) "Woodeaton", Journal Roman Studies, vii, 1917, pp.98-119

Taylor, M.V. (1963) "Statuettes Of Horsemen And Horses And Other Votive Objects From Brigstock, Northants.", Antiquaries Journal, XLIII, 1963, pp.264-268

Tebbutt, C.F. (1961) "Celtic Linch-pin Heads from Colne Fen, Huntingdonshire", Antiquaries Journal, XLI, 1961, pp.253-258

Tenney, F. (1959) An Economic Survey Of Ancient Rome, 5 vols., Paterson, New Jersey, 1959

Theilade, H. (1972) Crafts Unlimited Catalogue, 1972

Thomas, C. (1963) "The Interpretation of the Pictish Symbols", Archaeological Journal, cxx, 1963, pp.31-97

Thompson, F.H. (1954) "The Roman Aquaduct At Lincoln", Archaeological Journal, cxi, 1954, pp.106-128

Thompson, F.H. (1965) Roman Cheshire, Chester, 1965

Thurnan, J. (1871) "On Ancient British Barrows, especially those of Wiltshire and the adjoining counties", Archaeologia, 43, 1871, pp.285-544

Toll, N.P. (1949) "Enamelled Bronzes" pp.35-43 of Part IV, The Bronze Objects (by T.G. Finch & N.P. Toll), of the Final Report of The Excavations At Dura Europos (ed. by M.I. Rostovzeff et al.), New Haven, 1949

Trollope, E. (1857) Exhibition by, Archaeological Journal, xiv, 1857, p.75

Tylecote, H.F. (1962) Metallurgy In Archaeology, London, 1962

Vargin, V. (1967) Technology of Enamels, (ed. Shaw), London, 1967

Victoria County History (1901) The Victoria History Of The County Of Norfolk, vol. I, London, 1901

Wacher, J.S. (1969) Excavations at Brough-on-Humber 1958-61, Report Research Committee Society Antiquaries London no.XXV, Leeds, 1969

Wainwright, F.T. (1953) "A Souterrain Identified in Angus", Antiquaries Journal, XXXIII, 1953, pp.65-71

Wainwright, G. (1979) Gussage All Saints, London, 1979

Wainwright, G. & Spratling, M. (1973) "The Iron Age Settlement of Gussage All Saints", Antiquity, xlvii, 1973, pp.109-130

Ward Perkins, J.B. (1939) "Iron Age Metal Horse Bits of the British Isles", Proc. Prehistoric Society, V, 1939, pp.173-192

Ward Perkins, J.B. (1940) "Two Early Linch Pins from King's Langley, Herts., and from Tiddington, Stratford-on-Avon", Antiquaries Journal, XX, 1940, pp.358-387

Watson, W. (1949) "Belgic Bronzes And Pottery Found At Felmersham-On-Ouse, Bedfordshire", Antiquaries Journal, XXIX, 1949, pp.37-61

Webster, G. (1958) "The Roman Military Advance under Ostorius Scapula", *Archaeological Journal*, cxv. 1958, pp. 49-58

Webster, G. (1971) "A hoard of Roman military equipment from Fremington Hagg", pp. 107-126 of Butler (1971)

Webster, G. (1975) *The Cornovii*, London, 1975

Webster, G. (1978) Review of *Early Celtic Art in North Britain*, *Britannia*, IX, 1978, pp. 490-491

Webster, L.E. (1976) "Medieval Britain in 1975. Part I Pre-Conquest", *Medieval Archaeology*, 20, 1976, pp. 158-176

Webster, L.E. (1977) "Medieval Britain in 1976. Part 1 Pre-Conquest", *Medieval Archaeology*, 21, 1977, pp. 204-222

Wenger, (1967) *Wenger Metal Enamelling Handbook*, Stoke-on-Trent, 1957

Werner, A.E. & Organ, R. (1964) "Scientific report on an enamelled brooch from Wall, Staffs.", *Trans. Lichfield and S. Staffs. Arch. and Hist. Society*, 5, 1964, pp. 46.47

Westell, W.P. (1931) "Notes on a Romano-British Cemetery in Hertfordshire", *Proc. Society Antiquaries Scotland*, LXVI, 1931-2, pp. 105-113

Wheeler, R.E.M. (1923) "Segontium And The Roman Occupation Of Wales", *Y Cymmrodor*, xxxiii, 1923, pp. 1-186

Wheeler, R.E.M. & T.V. (1928) "The Roman Amphitheatre at Caerlson, Monmouthshire", *Archaeologia*, 78, 1928, pp. 111-218

Wheeler, R.E.M. & T.V. (1932) *Report on the Excavation of the Prehistoric, Roman and Post-Roman Site in Lydney Park, Gloucestershire*, Report Research Committee Society Antiquaries London no. IX, London, 1932

Wheeler, R.E.M. & T.V. (1936) *Verulamium, A Belgic and Two Roman Cities*, Report Research Committee Society Antiquaries London no. XI, London, 1936

Wheeler, R.E.M. (1943) *Maiden Castle, Dorset*, Report Research Committee Society Antiquaries London no. XII, Oxford, 1943

Wheeler, R.E.M. (1946) — London In Roman Times, London Museum Catalogues No. 3, London, 1946

Wild, J.P. (1970) — "Button-and-Loop Fasteners in the Roman Provinces", Britannia, I, 1970, pp. 137-155

Williams, A. (1948) — Roman Canterbury No. 3 An Account of the Excavations carried out during 1945, Medici Society, London, 1948

Wilson, D.M. (1964) — Anglo-Saxon Ornamental Metalwork 700-1100 In The British Museum, London, British Museum, 1964

Wilson, D.M. (1969) — "Medieval Britain in 1968. Part I Pre-Conquest" Medieval Archaeology, 13, 1969, pp. 230-243

Wilson, D.R. (1971) — "Roman Britain in 1970 I. Sites Explored", Britannia, II, 1971, pp. 243-288

Wilson, D.R. (1973) — "Roman Britain in 1972. I. Sites Explored", Britannia, IV, 1973, pp. 271-323

Wright, R.P. and Hassall, M.W.C. (1973) — "Roman Britain In 1972. II. Inscriptions", Britannia, IV, 1973, pp. 324-337

Yorkshire Museum, (1892) — Handbook of Antiquities, York, 1892

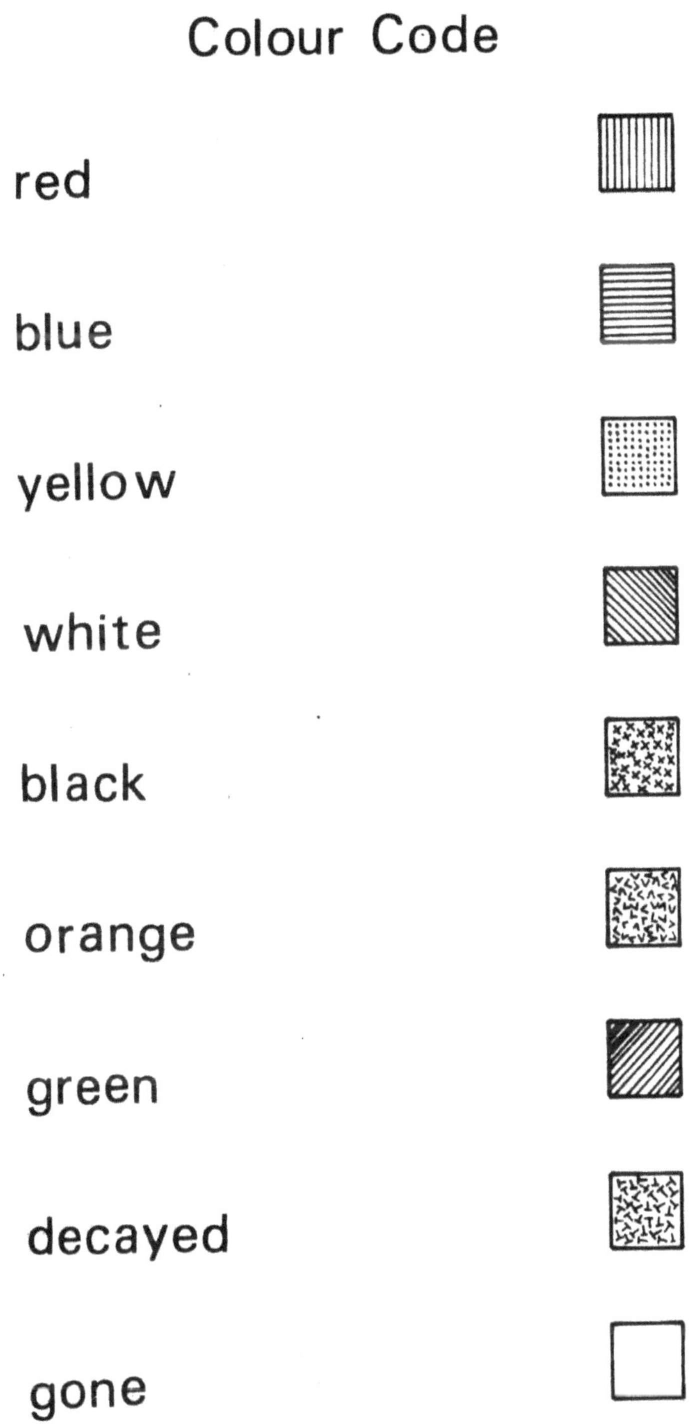
Colour Code
red
blue
yellow
white
black
orange
green
decayed
gone

FIGURE 1

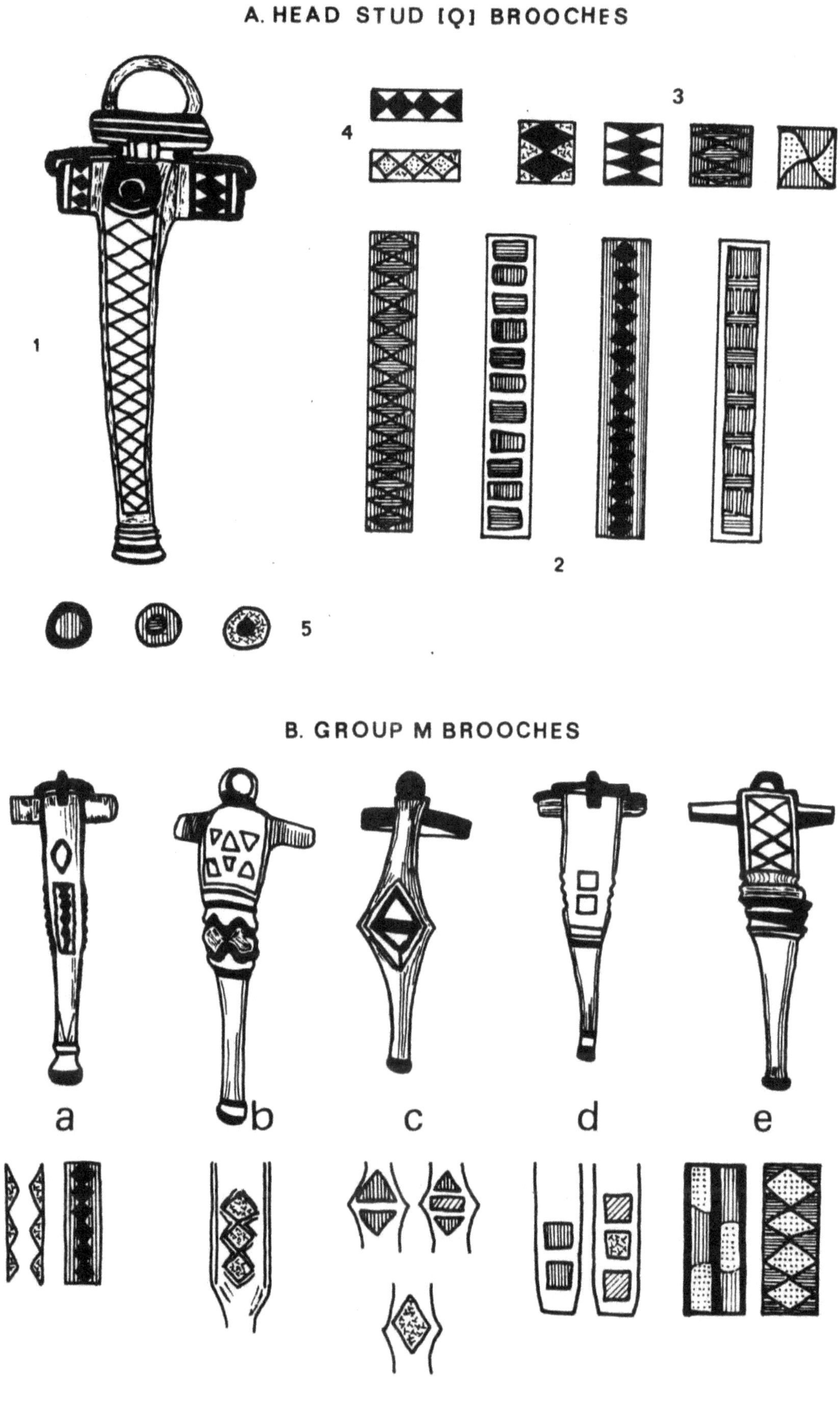

FIGURE 2

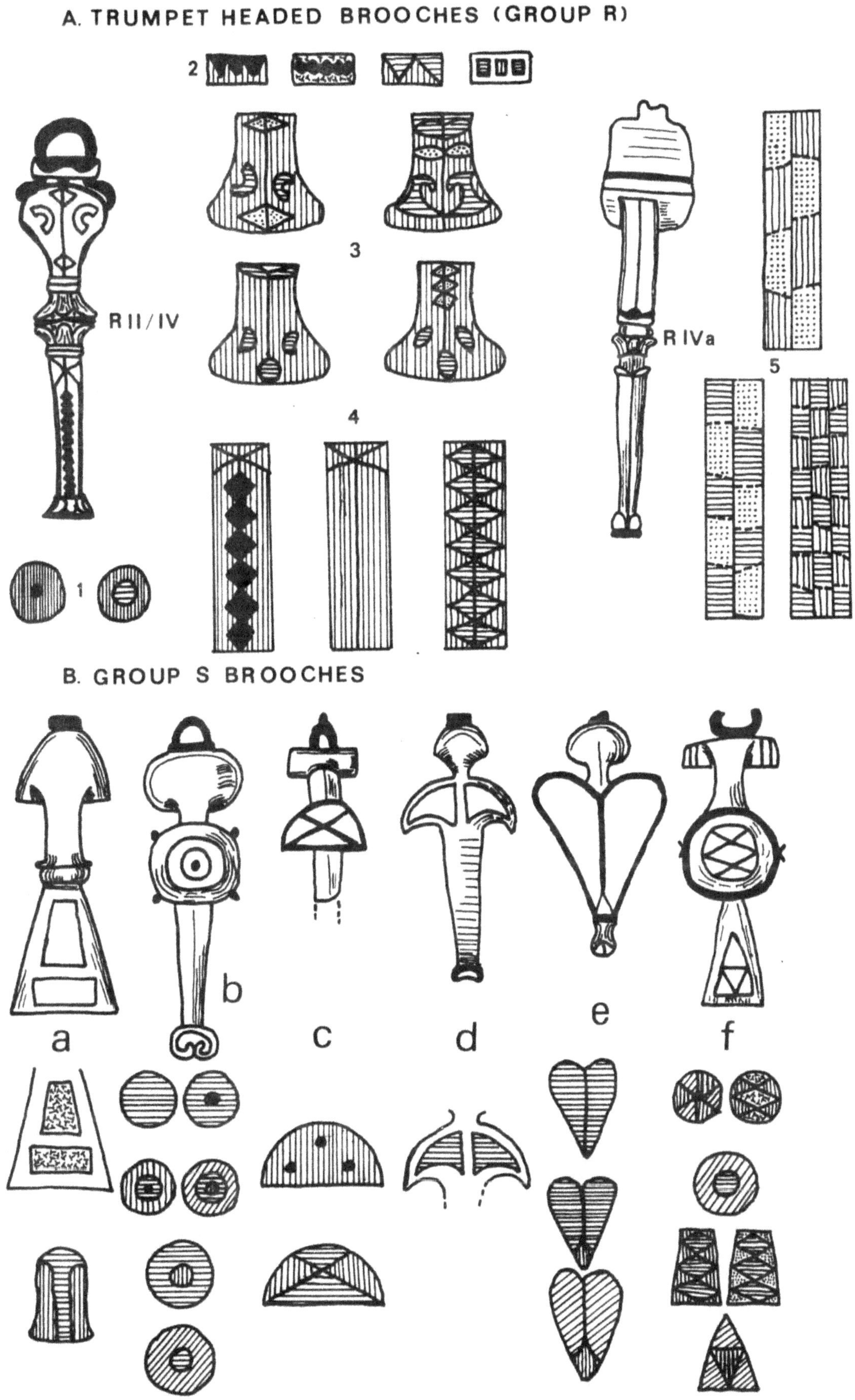

A. TRUMPET HEADED BROOCHES (GROUP R)
2
RII/IV
3
R IVa
5
4
1
B. GROUP S BROOCHES
a
b
c
d
e
f

FIGURE 3

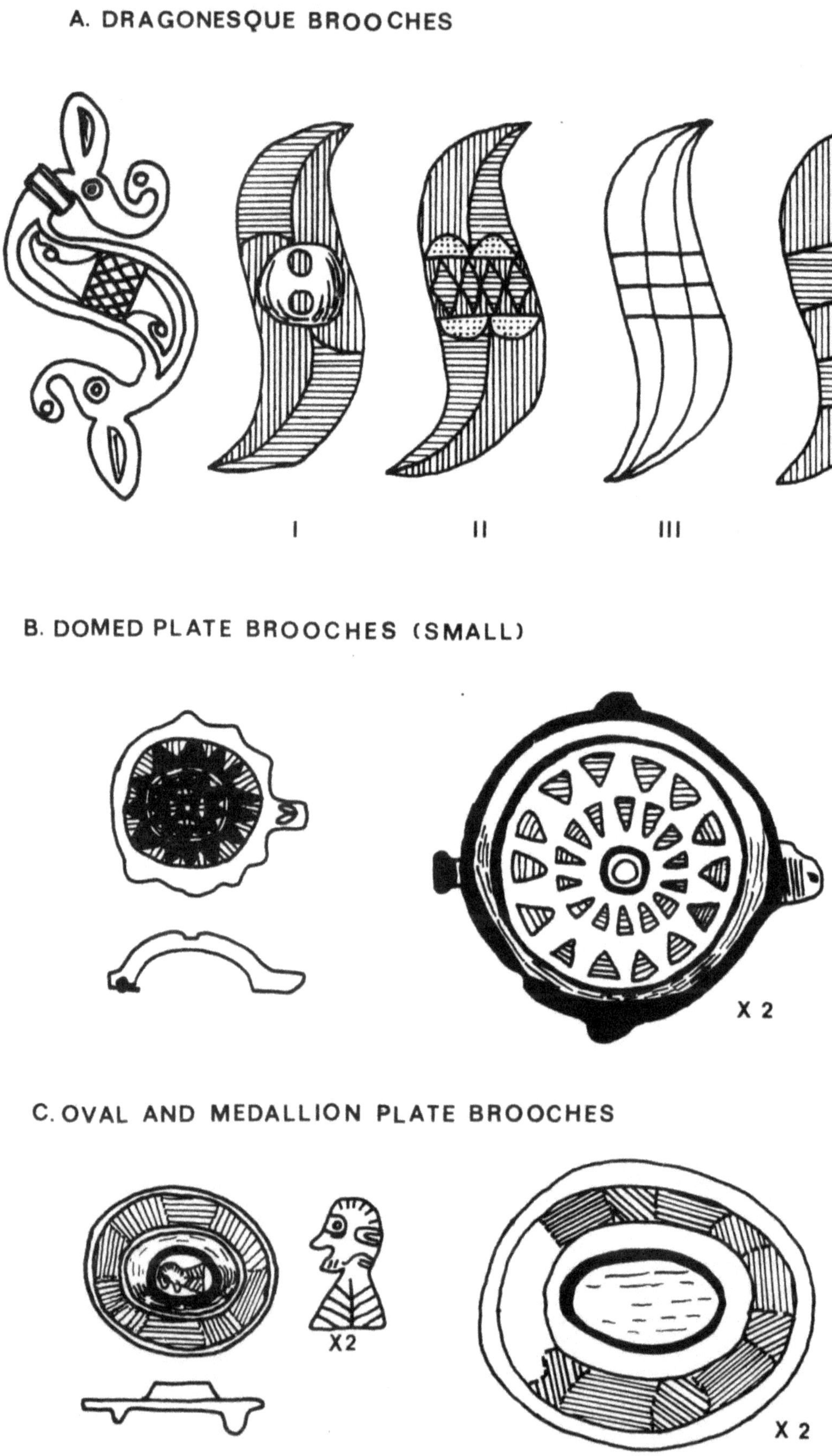
A. DRAGONESQUE BROOCHES
I
II
III
IV
B. DOMED PLATE BROOCHES (SMALL)
X 2
C. OVAL AND MEDALLION PLATE BROOCHES
X2
X 2

FIGURE 4

DISC BROOCHES

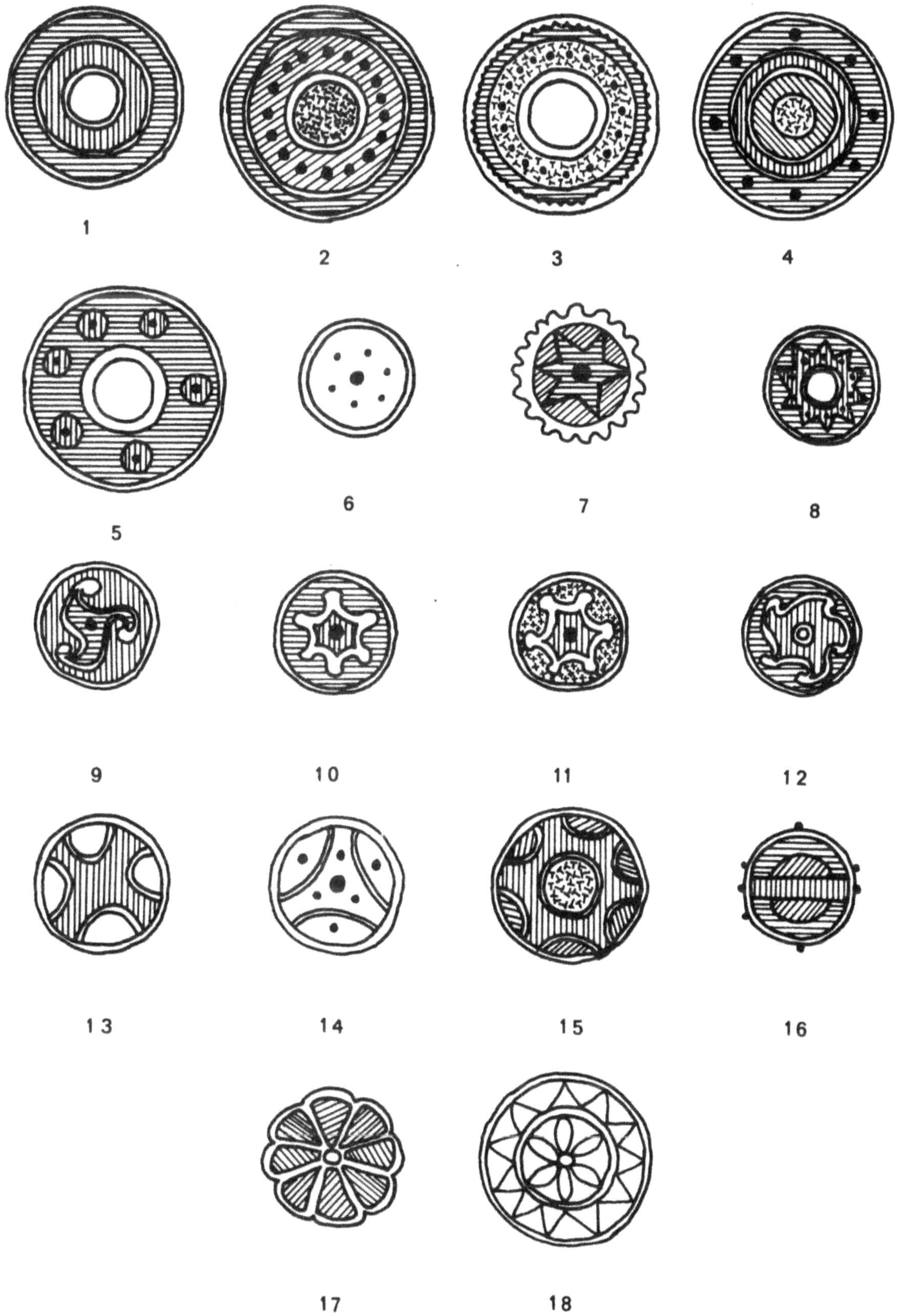

FIGURE 5

ANIMAL PLATE BROOCHES

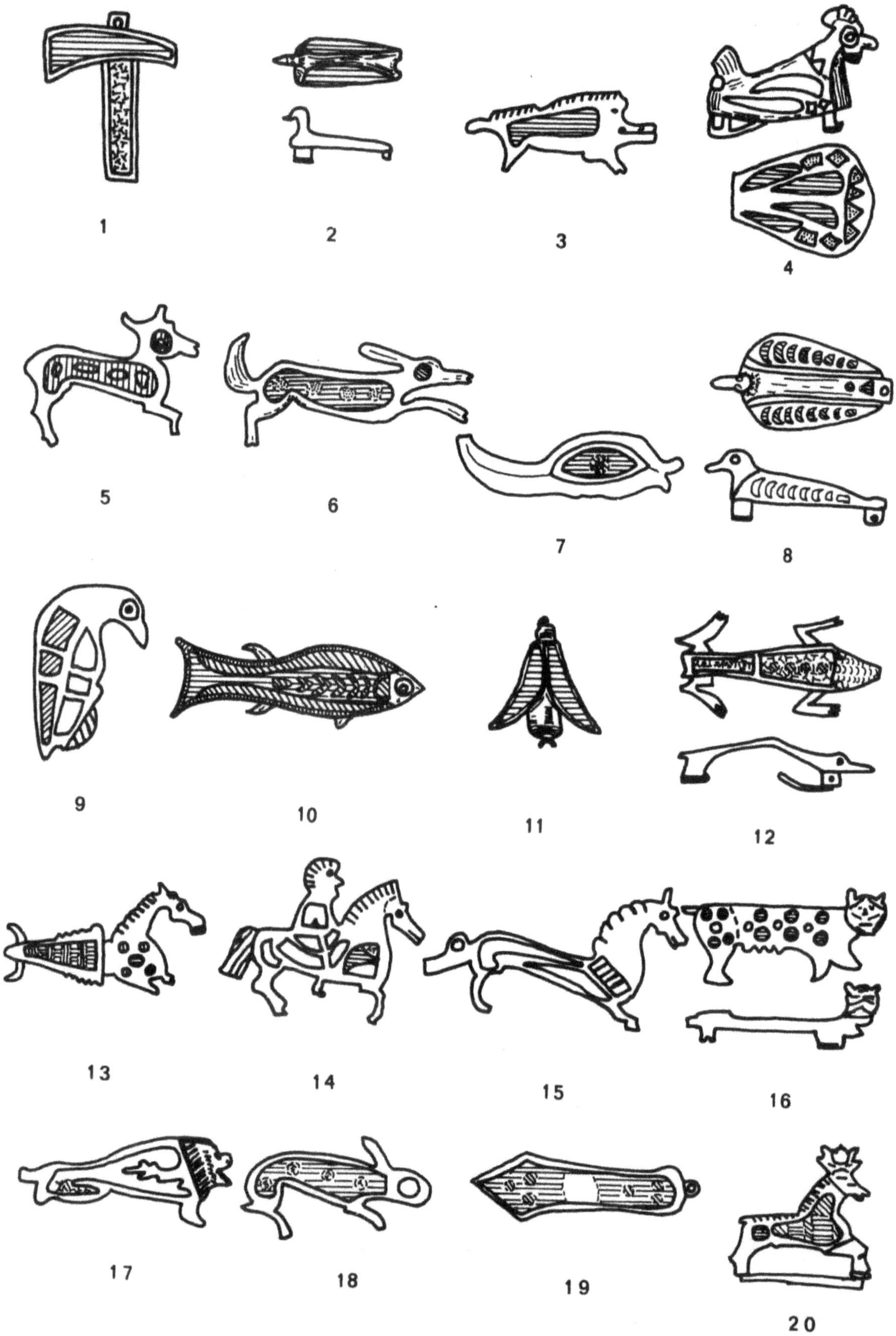

FIGURE 6

A. BUTTON AND LOOP FASTENERS

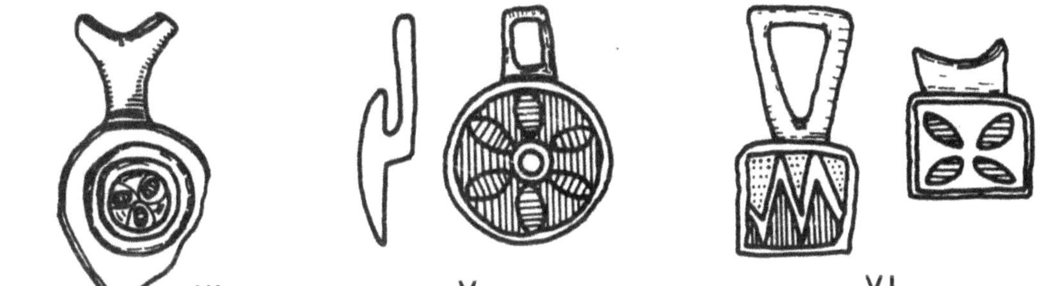

III V VI

B. FLAT CIRCULAR STUDS

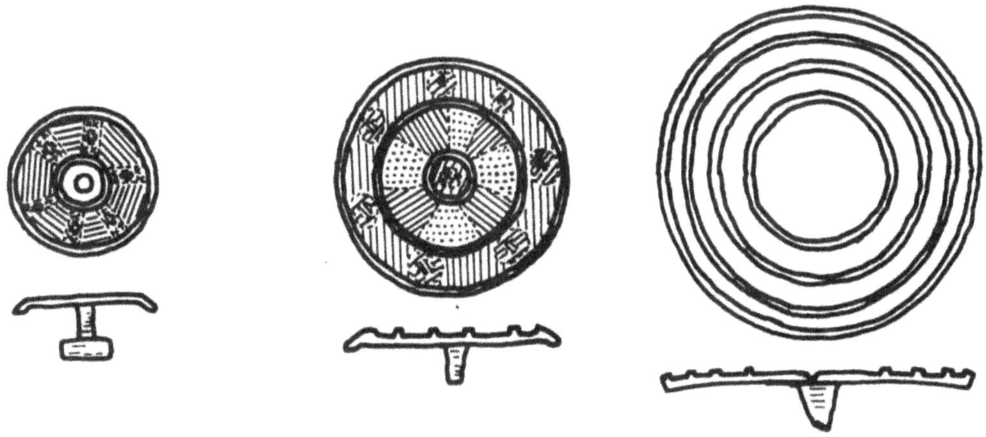

C. SEAL BOX LIDS

1 2 3 4 5

FIGURE 7

A. MINIATURE STANDS OR STOOLS

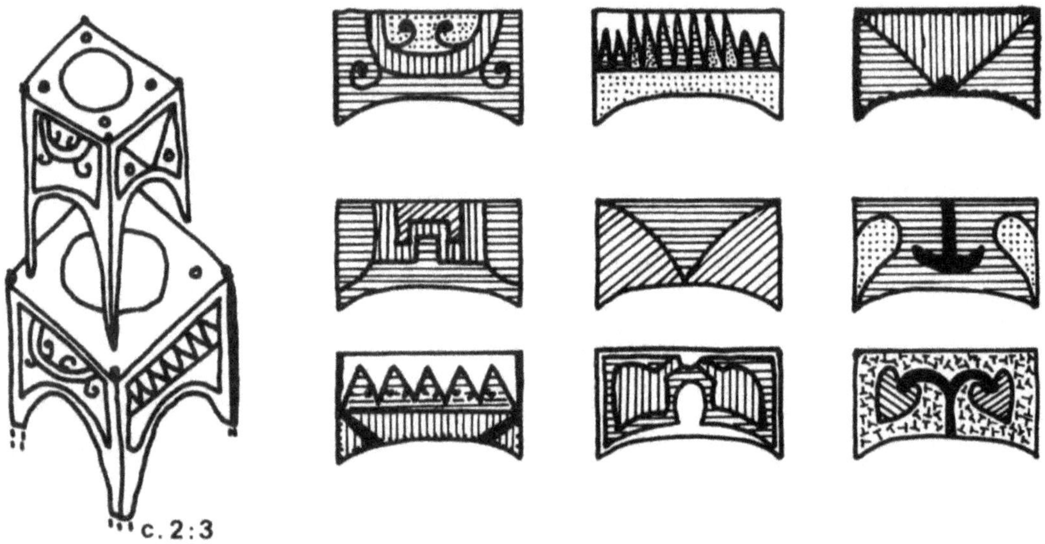

B. FINGER – RINGS

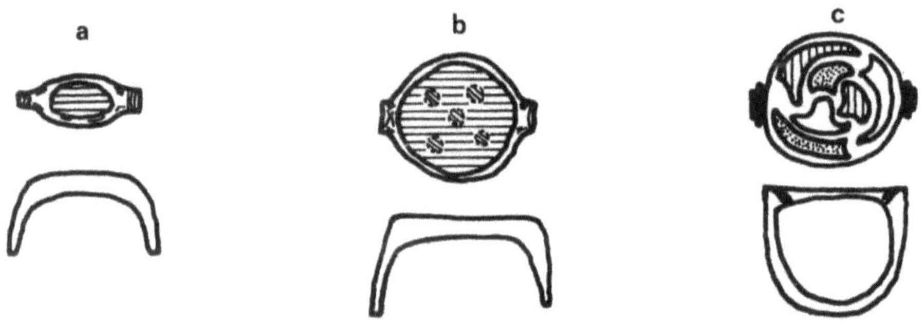

C. VESSELS

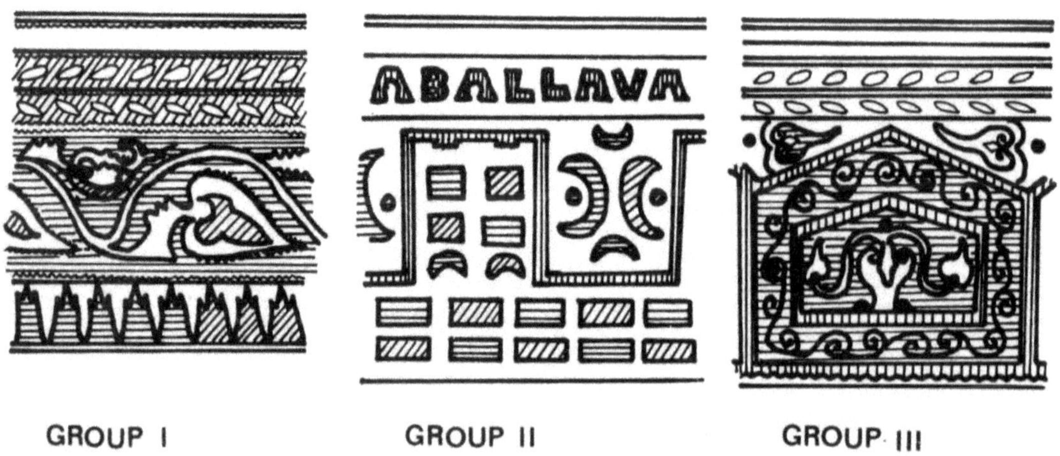

FIGURE 8

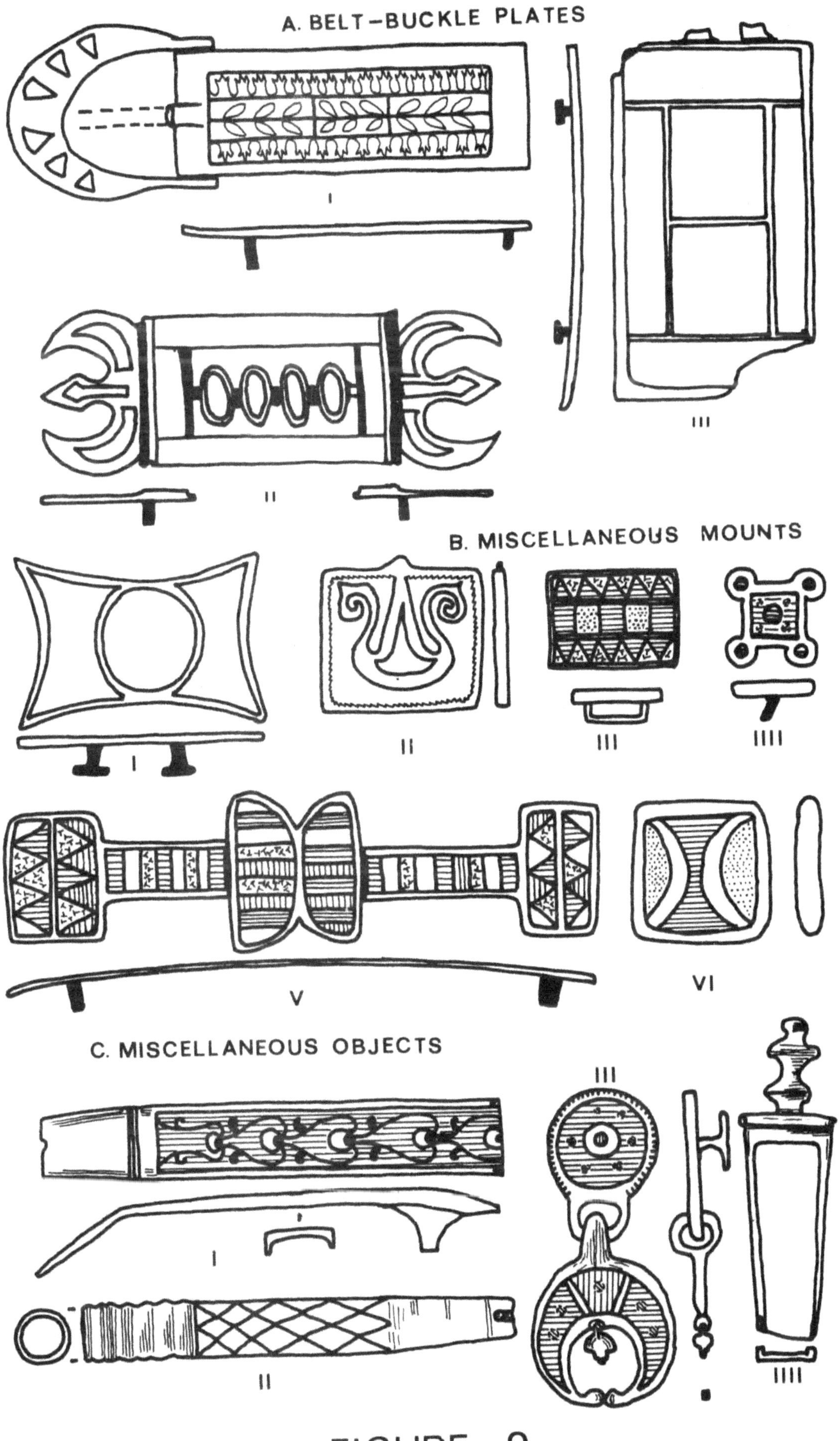
A. BELT–BUCKLE PLATES
I
II
III
B. MISCELLANEOUS MOUNTS
I
II
III
IIII
V
VI
C. MISCELLANEOUS OBJECTS
I
II
III
IIII

FIGURE 9

Techniques [Iron Age]

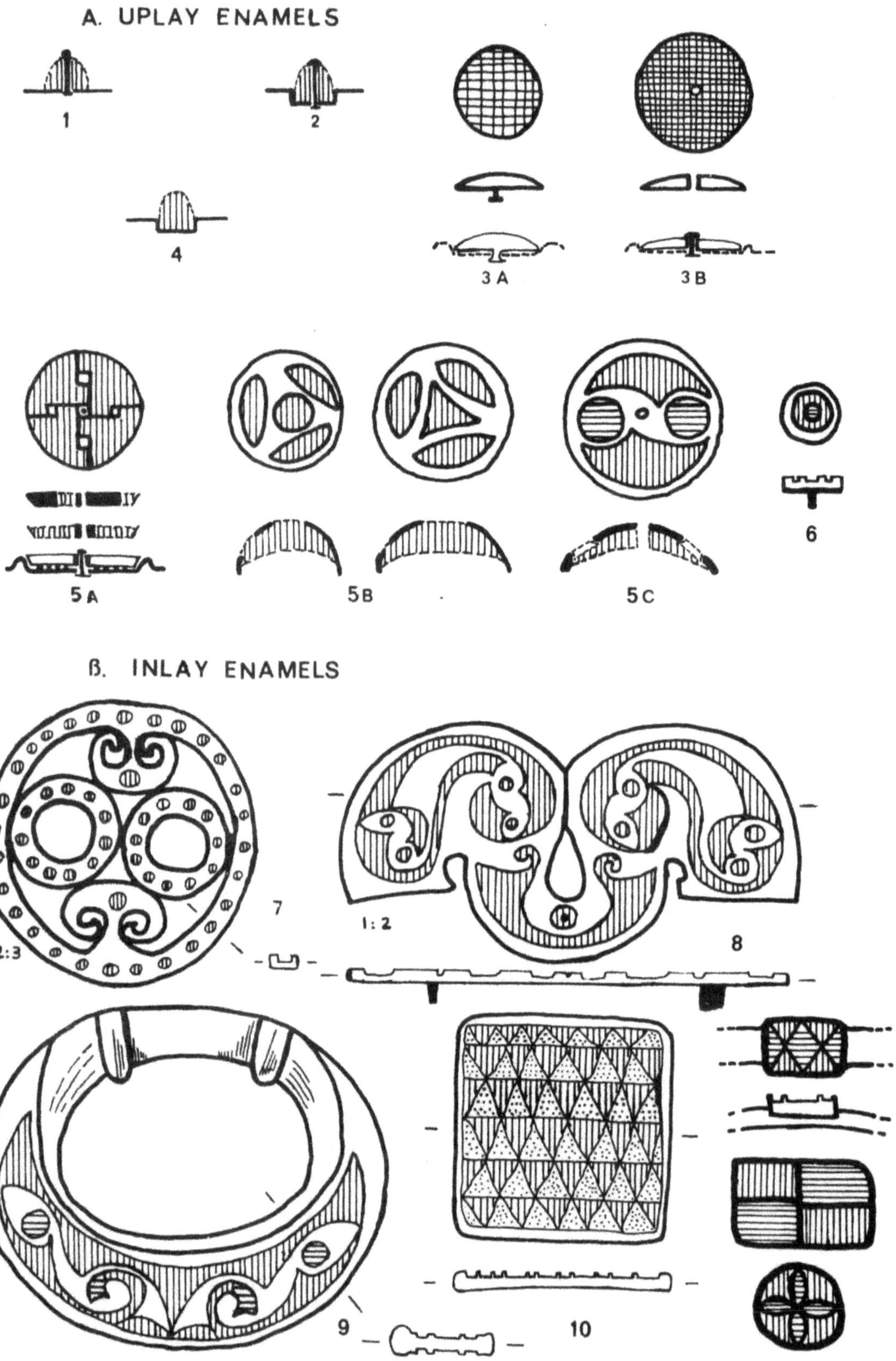

FIGURE 10

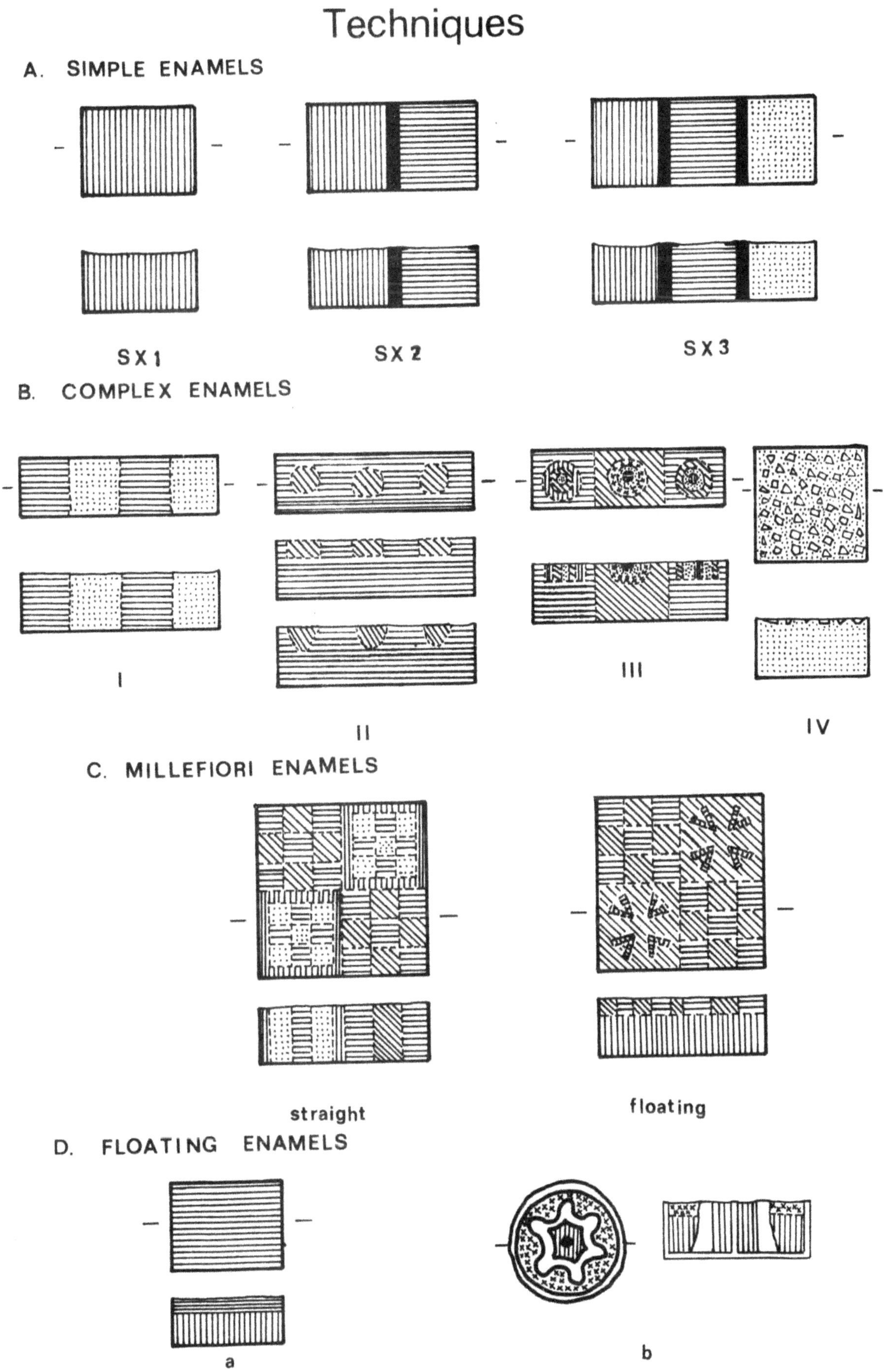
Techniques
A. SIMPLE ENAMELS
SX1
SX2
SX3
B. COMPLEX ENAMELS
I
II
III
IV
C. MILLEFIORI ENAMELS
straight
floating
D. FLOATING ENAMELS
a
b

FIGURE II

THE MILLEFIORI PATTERNS

TYPE A - basic pattern of nine squares

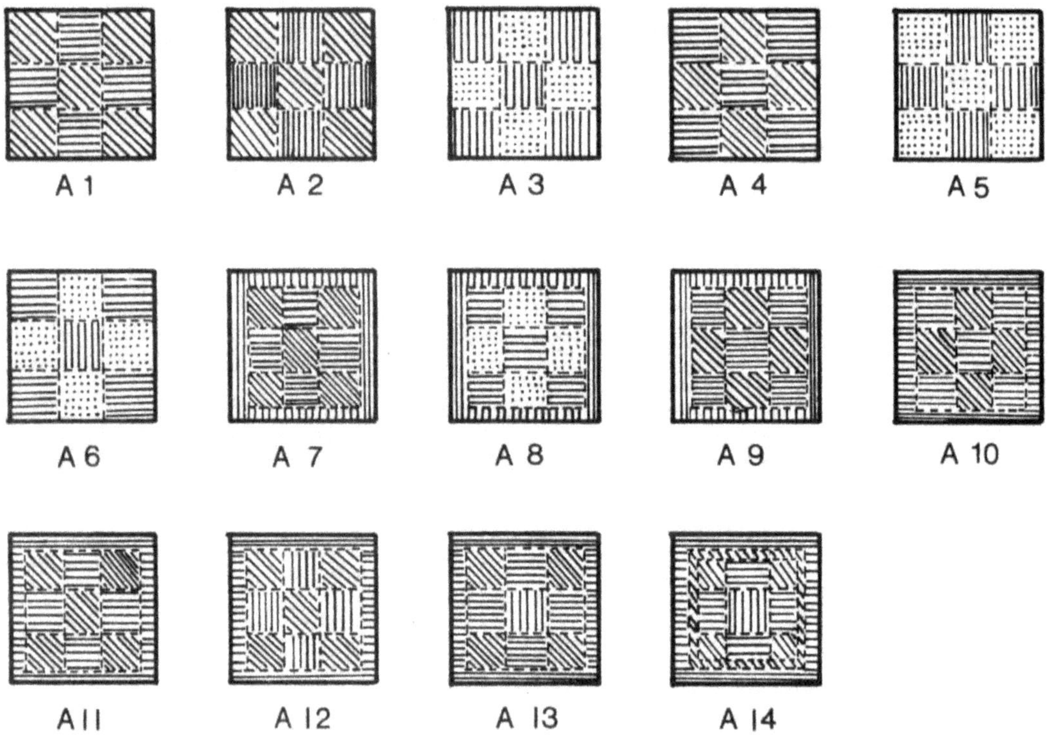

TYPE B - basic pattern of twenty-five squares

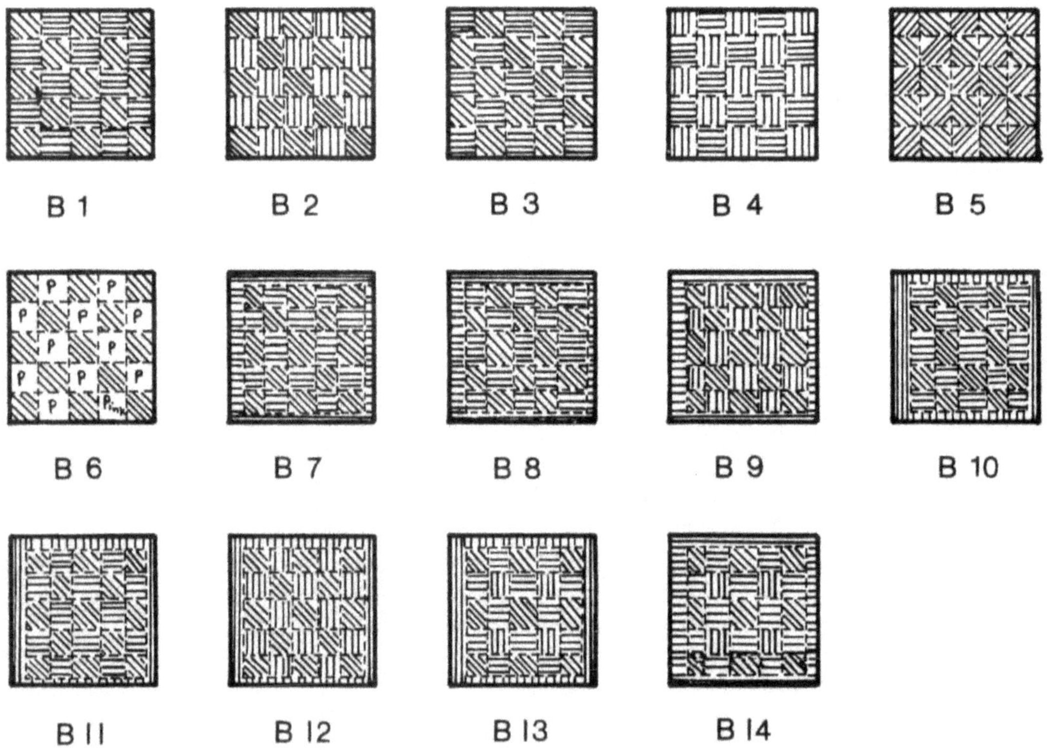

FIGURE 12

THE MILLEFIORI PATTERNS

TYPE C – basic pattern of ground with floweret

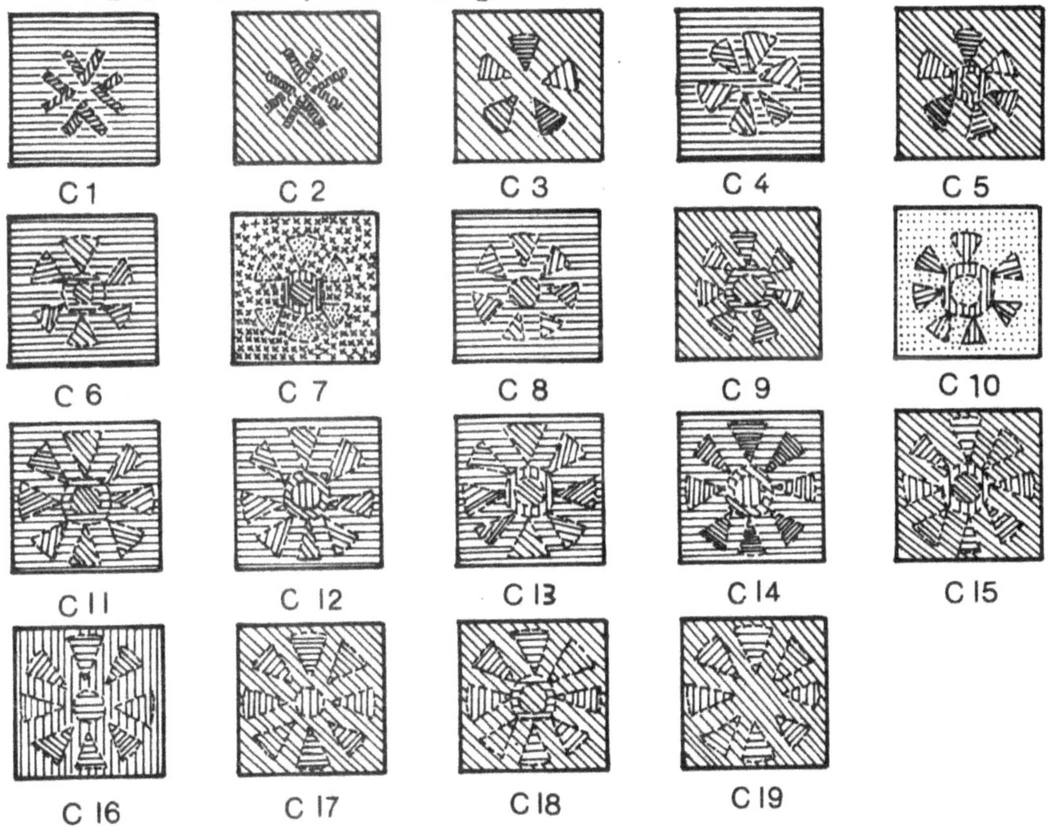

TYPE D – miscellaneous patterns

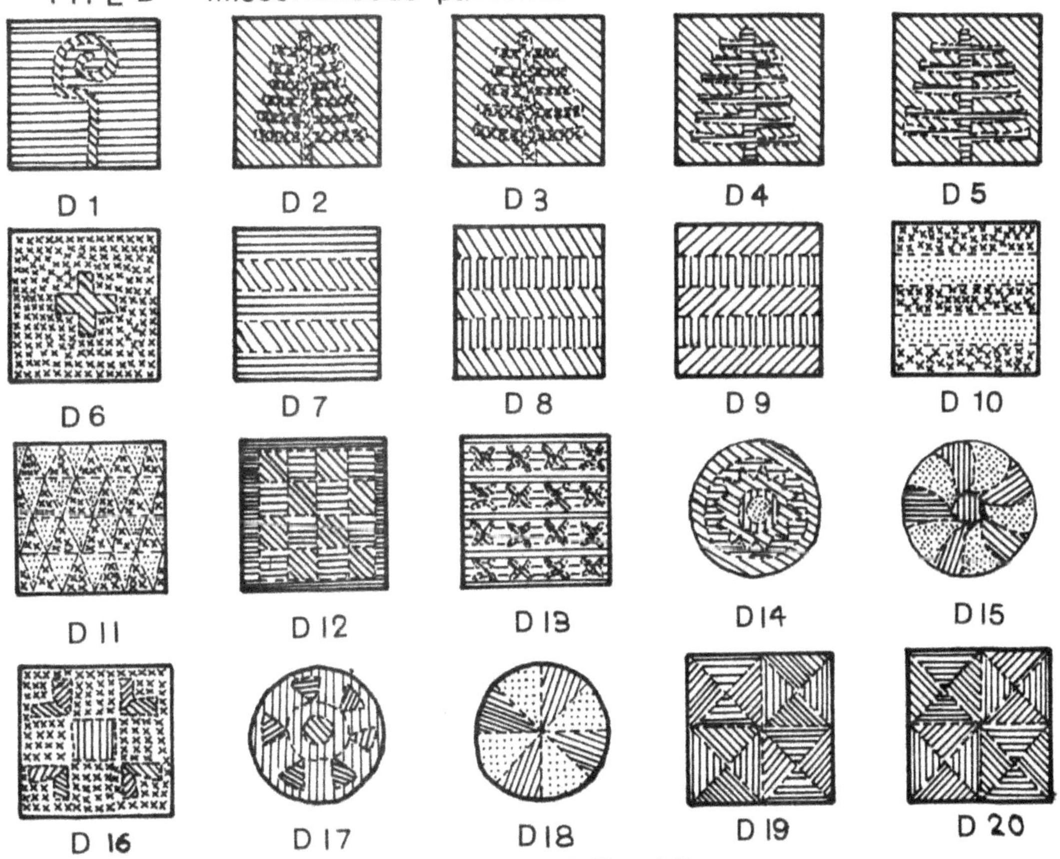

FIGURE 13

A. MODERN RAW MATERIALS FOR ENAMELLING

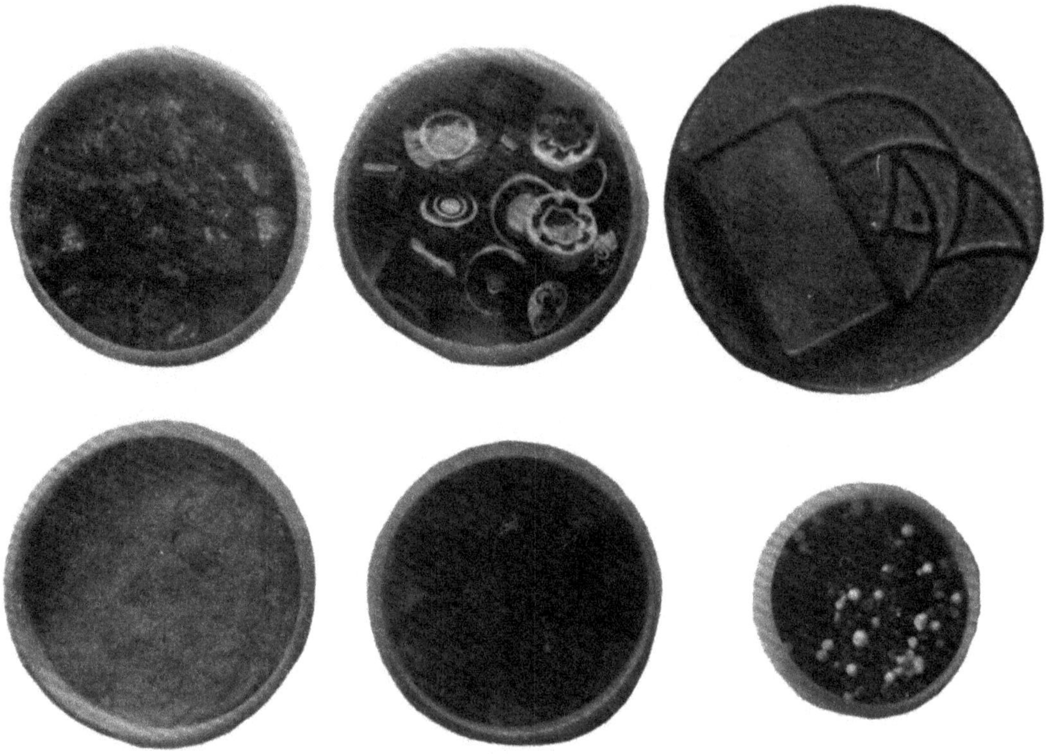

B. REPRODUCTIONS OF SIMPLE ENAMELS

FIGURE 14

A. REPRODUCTIONS OF COMPLEX ENAMELS

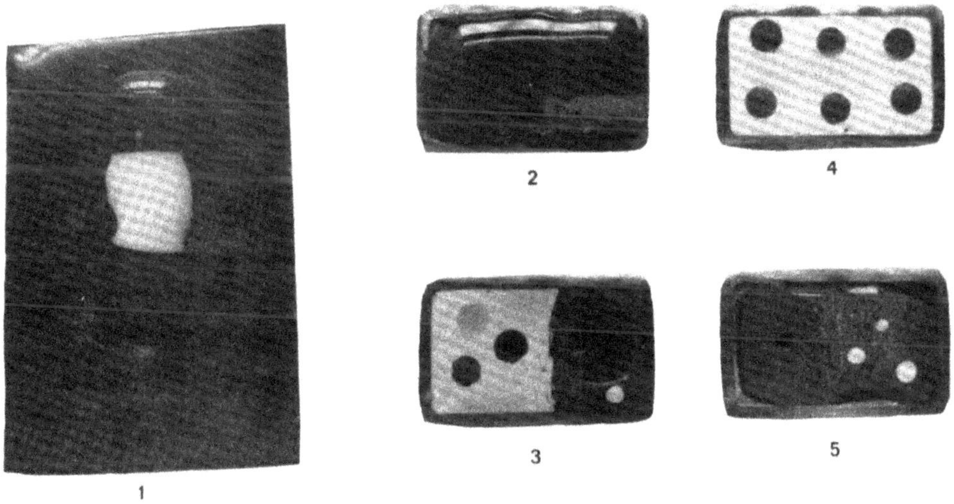

B. THE PROCESS OF MAKING A STRAIGHT MILLEFIORI ENAMEL

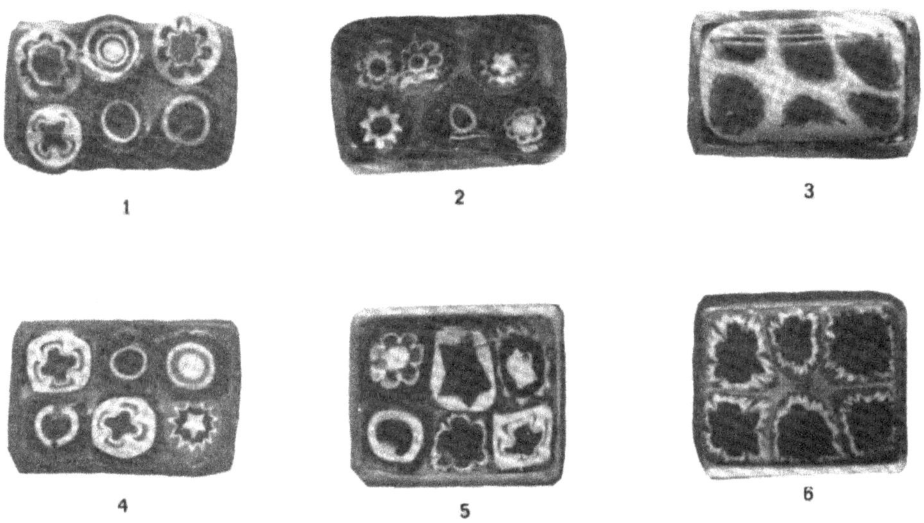

FIGURE 15

www.ingramcontent.com/pod-product-compliance
Lightning Source LLC
LaVergne TN
LVHW071210210926
848482LV00015B/771